American Book Company
The Standards Experts

Mastering the Georgia Mathematics III Course

Erica Day

Casey Dudek

Colleen Pintozzi

Bryan Roberts

Timothy Trowbridge

American Book Company
P. O. Box 2638
Woodstock, Georgia 30188-1383
Toll Free 1 (888) 264-5877 Phone (770) 928-2834
Toll Free Fax 1 (866) 827-3240
WEB SITE: www.americanbookcompany.com

Acknowledgements

In preparing this book, we would like to acknowledge Mary Stoddard and Eric Field for their contributions in developing graphics for this book and Camille Woodhouse for her contributions in formatting this book. We would also like to thank our many students whose needs and questions inspired us to write this text.

Printed in the United States of America
2/11

Contents

Contents

Preface

Mastering the Georgia Mathematics III Course will help you review and learn important concepts and skills related to algebra, geometry, and data analysis. First, take the Diagnostic Test beginning on page 1 of the book. Next, complete the evaluation chart with your instructor in order to help you identify the chapters which require your careful attention. When you have finished your review of all of the material your teacher assigns, take the practice test to evaluate your understanding of the material presented in this book. **The materials in this book are based on the Georgia Performance Standards that are published by the Georgia Department of Education. The complete list of standards is located in the Answer Key. Each question in the Diagnostic and Practice Tests is referenced to the standard, as is the beginning of each chapter.**

This book contains several sections. These sections are as follows: 1) A Diagnostic Test; 2) Chapters that teach the concepts and skills for *Mastering the Georgia Mathematics III Course*; and 3) Two Practice Tests. Answers to the tests and exercises are in a separate manual.

ABOUT THE AUTHORS

Erica Day has a Bachelor of Science Degree in Mathematics and is working on a Master of Science Degree in Mathematics. She graduated with high honors from Kennesaw State University in Kennesaw, Georgia. She has also tutored all levels of mathematics, ranging from high school algebra and geometry to university-level statistics, calculus, and linear algebra. She is currently writing and editing mathematics books for American Book Company, where she has coauthored numerous books, such as *Passing the Georgia Algebra I End of Course*, *Passing the Georgia High School Graduation Test in Mathematics*, *Passing the Arizona AIMS in Mathematics*, and *Passing the New Jersey HSPA in Mathematics*, to help students pass graduation and end of course exams.

Casey Dudek has taught math in both public and private high schools for seven years. He earned a Bachelor of Science degree in Math Education from the University of Georgia. He has helped write and edit several books, including *Passing the Louisiana Geometry End-of-Course Test* and *Mastering the iLEAP Math Test in Grade 9*.

Colleen Pintozzi has taught mathematics at the middle school, junior high, senior high, and adult level for 22 years. She holds a B.S. degree from Wright State University in Dayton, Ohio and has done graduate work at Wright State University, Duke University, and the University of North Carolina at Chapel Hill. She is the author of many mathematics books including such best-sellers as *Basics Made Easy: Mathematics Review, Passing the New Alabama Graduation Exam in Mathematics, Passing the Louisiana LEAP 21 GEE, Passing the Indiana ISTEP+ GQE in Mathematics, Passing the Minnesota Basic Standards Test in Mathematics,* and *Passing the Nevada High School Proficiency Exam in Mathematics.*

Bryan Roberts is working on a Bachelor of Science degree in Mathematics and Statistics at Kennesaw State University in Kennesaw, GA. He is currently a junior and has made the Dean's List multiple times during his undergraduate career. He is currently participating in an internship in mathematics and other departments for American Book Company where he does his editing and writing of books and also creates databases of questions for multiple grade levels. He is currently writing and editing mathematics books for American Book Company, where he has coauthored *Mastering the TCAP in 7th Grade Mathematics*.

Timothy Trowbridge graduated with summa cum laude honors from Hawaii Loa College with a Bachelor of Arts Degree in Mathematics. He taught in Japan as a participant in the Japan Exchange and Teaching (JET) Program, and he has written and edited parts of various mathematics textbooks for several major educational publishers.

Diagnostic Test

Part 1

1. What is the equation of a circle that has a center at $(-3, 5)$ and a diameter of 8?

   A. $(x - 3)^2 + (y + 5)^2 = 16$
   B. $(x + 3)^2 + (y - 5)^2 = 16$
   C. $(x - 3)^2 + (y + 5)^2 = 64$
   D. $(x + 3)^2 + (y - 5)^2 = 64$

   MM3G1a

2. What is the equation of a circle that has center at $(12, -7)$ and a radius of 13?

   A. $(x - 12)^2 + (y + 7)^2 = 169$
   B. $(x + 12)^2 + (y - 7)^2 = 169$
   C. $(x - 12)^2 + (y + 7)^2 = 13$
   D. $(x + 12)^2 + (y - 7)^2 = 13$

   MM3G1a

3. What is the equation of a circle that has center 7 units to the left of the origin and a radius of 11?

   A. $(x + 7)^2 + y^2 = 121$
   B. $(x - 7)^2 + y^2 = 11$
   C. $x^2 + (y - 7)^2 = 11$
   D. $(x - 7)^2 + y^2 = 121$

   MM3G1a

4. A circle has a radius of 5 units with its center at the origin. What is the equation of a line that is tangent to this circle at the point $\left(\dfrac{5\sqrt{2}}{2}, \dfrac{5\sqrt{2}}{2} \right)$?

   A. $f(x) = x + 5\sqrt{2}$
   B. $f(x) = -x + 5\sqrt{2}$
   C. $f(x) = -x - 5\sqrt{2}$
   D. $f(x) = x - 5\sqrt{2}$

   MM3G1c

5. A circle has a radius of $2\sqrt{13}$ units with its center at $(-2, 3)$. What is the equation of a line that is tangent to this circle at the point $(-6, 9)$?

   A. $f(x) = -\frac{2}{3}x + 13$
   B. $f(x) = \frac{2}{3}x - 13$
   C. $f(x) = -\frac{2}{3}x - 13$
   D. $f(x) = \frac{2}{3}x + 13$

   MM3G1c

6. A circle has a radius of 25 units with its center at $(21, 5)$. What is the equation of a line that is tangent to this circle at the point $(45, 12)$?

   A. $f(x) = -\frac{24}{7}x + 166\frac{2}{7}$
   B. $f(x) = -\frac{24}{7}x - 166\frac{2}{7}$
   C. $f(x) = \frac{24}{7}x + 166\frac{2}{7}$
   D. $f(x) = \frac{24}{7}x - 166\frac{2}{7}$

   MM3G1c

7. How many times do the two circles that have equations $(x - 10)^2 + y^2 = 16$ and $4(0.5x - 5)^2 + y^2 = 16$ intersect, if any?

   A. The two circles intersect at one point.
   B. The two circles intersect at two points.
   C. The two circles intersect at infinitely many points.
   D. The two circles do not intersect.

   MM3G1e

8. If the equation to Circle A is $x^2 + y^2 = 36$ and the equation to Circle B is $(x - 3)^2 + y^2 = 36$, how many points do the two circles intersect, if any?

   A. 0
   B. 1
   C. 2
   D. ∞

   MM3G1e

9. What does the graph of a circle with the equation $4x^2 + 4y^2 = 144$ look like?

A.

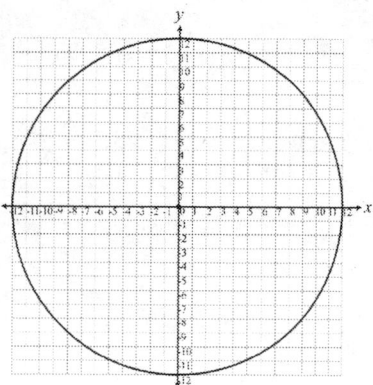

B.

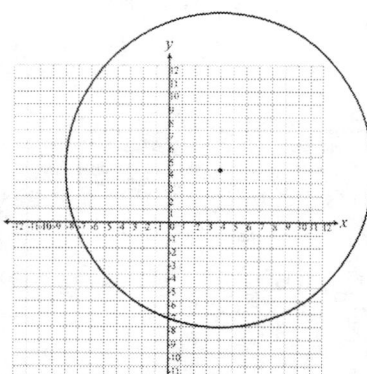

C.

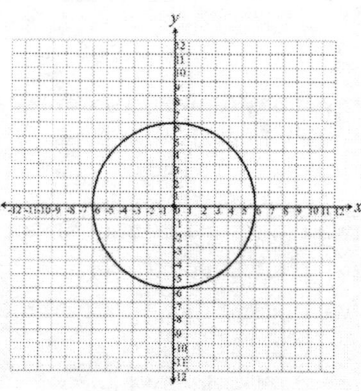

D.

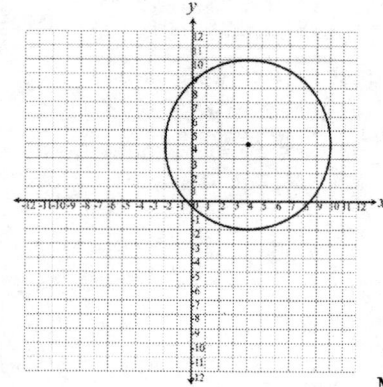

MM3G1b

10. What does the graph of a circle with the equation $2x^2 + 2y^2 = 8$ look like?

A.

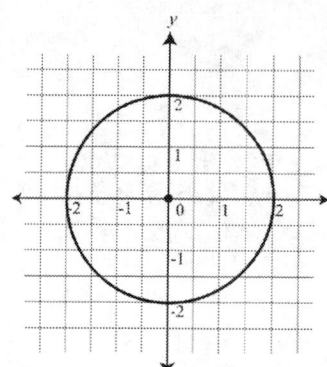

B.

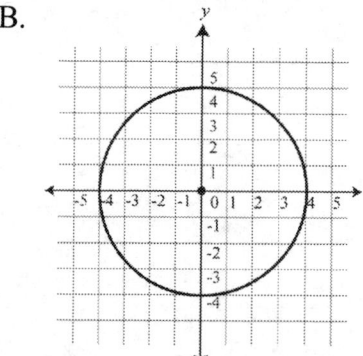

C.

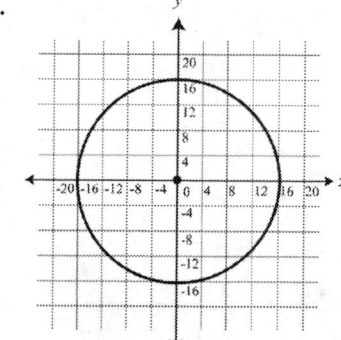

D.

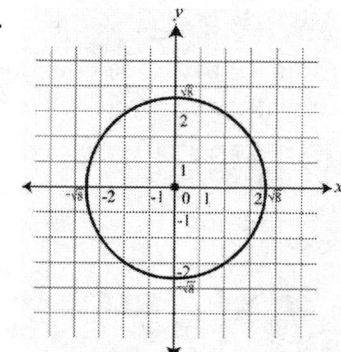

MM3G1b

2

11. There are two points in the 3-dimensional graph. Point A is located at $(-1, -1, -1)$, and Point B is located at $(2, 3, 4)$. What is the distance between the two points?

A. $5\sqrt{2}$
B. $2\sqrt{3}$
C. $\sqrt{6}$
D. 50

MM3G3b

12. What is the distance between two vertices of a rectangular prism if the two points are plotted at $(-3, 1, -4)$ and $(1, 2, 6)$?

A. $\sqrt{15}$
B. 3
C. $3\sqrt{13}$
D. $\sqrt{5}$

MM3G3b

13. What is the equation of a sphere whose center is located at $(4, 0, 5)$ with a radius that measures 19 units.

A. $(x - 4)^2 + (z - 5)^2 = 361$
B. $(x - 4)^2 + y^2 + (z - 5)^2 = 19$
C. $(x - 4)^2 + (z - 5)^2 = 19$
D. $(x - 4)^2 + y^2 + (z - 5)^2 = 361$

MM3G3c

14. An equation of a plane is $8x + 9y + 24z = 72$. Where does the plane cross the x-axis?

A. $(0, 8, 0)$
B. $(0, 0, 3)$
C. $(9, 0, 0)$
D. cannot be determined

MM3G3c

15. Find the zeros of the parabola by completing the square to the equation $y - x^2 + 8x = 10$.

A. $y = 4 \pm \sqrt{6}$
B. $x = 4 \pm \sqrt{6}$
C. $x = -4 \pm \sqrt{6}$
D. $y = -4 \pm \sqrt{6}$

MM3G2a

16. Classify the conic that has an equation $8x^2 - 48x + 18y^2 = 0$.

A. Circle
B. Hyperbola
C. Ellipse
D. Parabola

MM3G2a

17. Convert the equation of the ellipse to standard form.
$$16x^2 - 96x + 25y^2 + 100y = 156$$

A. $\dfrac{(x + 3)^2}{25} + \dfrac{(y - 2)^2}{16} = 1$

B. $\dfrac{(x + 3)^2}{16} + \dfrac{(y - 2)^2}{25} = 1$

C. $\dfrac{(x - 3)^2}{25} + \dfrac{(y + 2)^2}{16} = 1$

D. $\dfrac{(x - 3)^2}{16} + \dfrac{(y + 2)^2}{25} = 1$

MM3G2a

18. What is the equation of a horizontal ellipse that has a center at $(-3, 1)$ and $a = 3$ and $b = 7$?

A. $\dfrac{(x + 3)^2}{49} + \dfrac{(y - 1)^2}{9} = 1$

B. $\dfrac{(x - 3)^2}{7} + \dfrac{(y + 1)^2}{3} = 1$

C. $\dfrac{(x + 3)^2}{9} + \dfrac{(y - 1)^2}{49} = 1$

D. $\dfrac{(x + 3)^2}{7} + \dfrac{(y - 1)^2}{3} = 1$

MM3G2c

19. What is the equation of a circle that has a center at $(2, 3)$ and its area is 49π units?

A. $(x + 2)^2 + (y + 3)^2 = 49$
B. $(x - 3)^2 + (y - 2)^2 = 7$
C. $(x - 2)^2 + (y - 3)^2 = 49$
D. $(x + 2)^2 + (y + 3)^2 = 7$

MM3G2c

20. A hyperbola has an equation of $\dfrac{y^2}{16} - \dfrac{x^2}{9} = 1$. What are the coordinates for the foci?

   A. $(5, 0)$ and $(-5, 0)$

   B. $(0, 5)$ and $(0, -5)$

   C. $(25, 0)$ and $(-25, 0)$

   D. $(0, 25)$ and $(0, -25)$

MM3G2b

21. A hyperbola has an equation of $\dfrac{y^2}{9} - \dfrac{x^2}{16} = 1$. What are the equations for the asymptotes?

   A. $y = \pm\dfrac{3}{4}x$

   B. $y = \pm\dfrac{4}{3}x$

   C. $y = \pm\dfrac{9}{16}x$

   D. $y = \pm\dfrac{16}{9}x$

MM3G2b

22. What is the length of a diagonal in a rectangular prism that has endpoints with coordinates of $(0, 2, 3)$ and $(3, -1, 0)$?

   A. $9\sqrt{3}$
   B. 27
   C. $3\sqrt{3}$
   D. $3\sqrt{9}$

MM3G3a

23. What is the length of a diagonal in a rectangular prism that has endpoints with coordinates of $\left(2\sqrt{15}, 4\sqrt{3}, 2\right)$ and $\left(\sqrt{15}, 6\sqrt{3}, -1\right)$?

   A. 6
   B. 36
   C. -6
   D. -36

MM3G3a

24. If the equation of a circle is $(x - 2)^2 + (y + 3)^2 = 16$ and the equation of a line is $y = x - 3$, what are the point(s) of intersection, if any?

   A. $\left(1 + \sqrt{7}, -2 + \sqrt{7}\right)$
   $\left(1 - \sqrt{7}, -2 - \sqrt{7}\right)$

   B. $\left(1 - \sqrt{7}, -2 + \sqrt{7}\right)$
   $\left(1 + \sqrt{7}, -2 - \sqrt{7}\right)$

   C. $\left(1 + \sqrt{7}, -2 + \sqrt{7}\right)$
   $\left(1 + \sqrt{7}, 2 + \sqrt{7}\right)$

   D. no intersection

MM3G1d

25. If the equation of a circle is $(x + 4)^2 + (y - 2)^2 = 4$ and the equation of a line is $y = 2x + 2$, what are the point(s) of intersection, if any?

   A. $\left(\dfrac{-4 - 2\sqrt{11}}{5}, \dfrac{2 - 4\sqrt{11}}{5}\right)$
   $\left(\dfrac{-4 + 2\sqrt{11}}{5}, \dfrac{-2 + 4\sqrt{11}}{5}\right)$

   B. $\left(\dfrac{4 - 2\sqrt{11}}{5}, \dfrac{2 + 4\sqrt{11}}{5}\right)$
   $\left(\dfrac{-4 - 2\sqrt{11}}{5}, \dfrac{-2 + 4\sqrt{11}}{5}\right)$

   C. $\left(\dfrac{-4 + 2\sqrt{11}}{5}, \dfrac{-2 - 4\sqrt{11}}{5}\right)$
   $\left(\dfrac{4 - 2\sqrt{11}}{5}, \dfrac{2 + 4\sqrt{11}}{5}\right)$

   D. no intersection

MM3G1d

4

26. If $\sqrt[9]{17^4} = 17^x$, what is the value of x?

   A. $\frac{4}{9}$

   B. $\frac{2}{3}$

   C. $\frac{3}{2}$

   D. $\frac{9}{4}$

                    MM3A2a

27. Which of these expressions is equal to $10x^4 + 7x^5$?

   A. $6x^4 - 3x^5 - 4x^4 + 10x^5$

   B. $6x^4 - 3x^5 + 4x^4 + 10x^5$

   C. $6x^5 - 3x^4 - 4x^5 + 10x^4$

   D. $6x^5 - 3x^4 + 4x^5 + 10x^4$

                    MM3A2b

28. Which of these expressions is equal to -4?

   A. $2\log_2 3 - \log_2 144$

   B. $2\log_2 3 + \log_2 144$

   C. $3\log_2 2 - \log_2 144$

   D. $3\log_2 2 + \log_2 144$

                    MM3A2d

29. Which of these points is an intercept of the function $y = \left(\frac{1}{10}\right)^x$?

   A. $(0, 0)$

   B. $(0, 1)$

   C. $(1, 0)$

   D. $(1, 1)$

                    MM3A2e

30. Tim deposited $1,500 into a savings account offering continuously compounded interest. For 3 years, the interest rate stayed the same, and Tim didn't make any additional deposits or withdrawals. If his balance after 3 years was $1,800, what is its doubling time from the original deposit? Round to the nearest year.

   A. 5 years

   B. 8 years

   C. 11 years

   D. 14 years

                    MM3A2g

31. Which of these is a possible number of complex roots of the equation $x^5 - x^4 + 13x^3 - 13x^2 + 36x - 36 = 0$?

   A. 1

   B. 3

   C. 4

   D. 5

                    MM3A3a

32. To solve the equation $x^3 + 3x^2 = 9x + 27$, the functions $f(x) = x^3 + 3x^2$ and $g(x) = 9x + 27$ were graphed. At how many points do the graphs of the functions intersect?

   A. 0

   B. 1

   C. 2

   D. 3

                    MM3A3b

33. The graphs of the functions $f(x) = \left(\frac{1}{8}\right)(2^x) - 8$ and $g(x) = 8$ intersect at the point $(7, 8)$. What is the solution to the equation $\left(\frac{1}{8}\right)(2^x) - 8 = 8$?

   A. $x = -8$

   B. $x = -7$

   C. $x = 7$

   D. $x = 8$

                    MM3A3d

34. To solve the equation $\log_3(x^2 - 4x - 12) = 2$, the left side of the equation was rewritten so that it could be graphed on a graphing calculator, and then both sides of the equation were graphed on the same coordinate grid. At how many points do the two graphs intersect?

   A. 0

   B. 1

   C. 2

   D. 3

                    MM3A3b

35. The solution to an inequality written in interval notation is $[-4, 4] \cup [8, \infty)$. Which of these values does not satisfy the inequality?

A. -10
B. -4
C. 2
D. 8

MM3A3c

36. To solve the inequality $x^3 + 3x^2 < 16x + 48$, the functions $f(x) = x^3 + 3x^2$ and $g(x) = 16x + 48$ were graphed on the same coordinate grid. Which of these regions of the coordinate grid satisfy the inequality?

A. Those where $f(x)$ is above $g(x)$.
B. Those where $f(x)$ is at or above $g(x)$.
C. Those where $f(x)$ is below $g(x)$.
D. Those where $f(x)$ is at or below $g(x)$.

MM3A3b

Part 2

37. To solve the inequality $\left(\frac{1}{4}\right)(10^x) - 2 < 23$, which of these pairs of functions could be graphed?

    A. $f(x) = \left(\frac{1}{4}\right)(10^x) - 2$ and $g(x) = -23$

    B. $f(x) = \left(\frac{1}{4}\right)(10^x) - 2$ and $g(x) = 23$

    C. $f(x) = \left(\frac{1}{4}\right)(10^x) + 2$ and $g(x) = -23$

    D. $f(x) = \left(\frac{1}{4}\right)(10^x) + 2$ and $g(x) = 23$

    MM3A3d

38. The graphs of the functions $f(x) = \log(x - 0.2)$ and $g(x) = -1$ are shown below.

    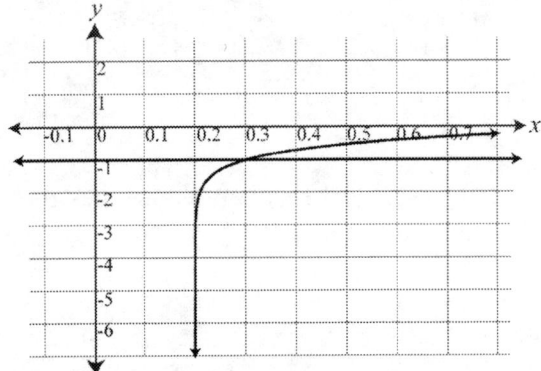

    What is the solution to the inequality $\log(x - 0.2) < -1$ in interval notation?

    A. $(0.2, 0.3)$
    B. $[0.2, 0.3)$
    C. $(0.2, 0.3]$
    D. $[0.2, 0.3]$

    MM3A3c

39. The end behavior of the graph of the polynomial function f is $f(x) \to \infty$ as $x \to -\infty$ and $f(x) \to -\infty$ as $x \to \infty$. As the value of x gets extremely large, what happens to the graph?

    A. It extends infinitely downward.
    B. It extends infinitely upward.
    C. It approaches the x-axis.
    D. It approaches the y-axis.

    MM3A1d

40. The graph of the function $f(x) = 64x^6$ was transformed to produce the graph of the function $f(x) = 64(x + 5)^6 - 3$. Which of these transformation were applied?

    A. a translation left and a translation down
    B. a translation left and a translation up
    C. a translation right and a translation down
    D. a translation right and a translation up

    MM3A1a

41. What is the end behavior of the graph of a polynomial function with an odd degree and a positive lead coefficient?

    A. $f(x) \to \infty$ as $x \to \pm\infty$

    B. $f(x) \to -\infty$ as $x \to \pm\infty$

    C. $f(x) \to -\infty$ as $x \to -\infty$ and $f(x) \to \infty$ as $x \to \infty$

    D. $f(x) \to \infty$ as $x \to -\infty$ and $f(x) \to -\infty$ as $x \to \infty$

    MM3A1d

42. Which of these statements accurately describes the function graphed below?

    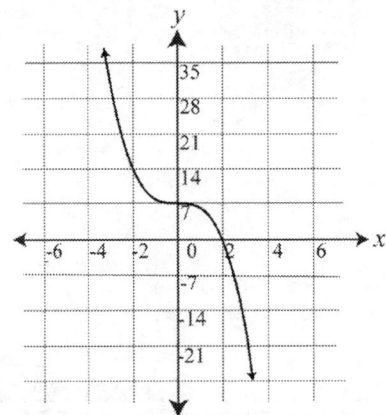

    A. It is neither even or odd.
    B. It is even.
    C. It is odd.
    D. It is both even and odd.

    MM3A1c

43. Consider the following equations:

$f(x) = 6x + 2$ and $f(x) = 3x + 2$

Which of the following statements is true concerning the graphs of these equations?

A. The lines are collinear.
B. The lines intersect at exactly one point.
C. The lines are parallel to each other.
D. The graphs of the lines intersect each other at the point $(2, 2)$.

MM3A5c

44. Which matrix represents the routes of this network?

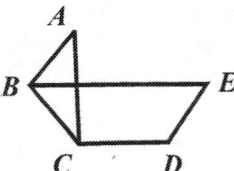

A.
$$
\begin{array}{c|ccccc}
 & A & B & C & D & E \\
\hline
A & 1 & 0 & 0 & 1 & 1 \\
B & 0 & 1 & 0 & 1 & 0 \\
C & 0 & 0 & 1 & 0 & 1 \\
D & 1 & 1 & 0 & 1 & 0 \\
E & 1 & 0 & 1 & 0 & 1 \\
\end{array}
$$

B.
$$
\begin{array}{c|ccccc}
 & A & B & C & D & E \\
\hline
A & 1 & 1 & 1 & 0 & 0 \\
B & 1 & 1 & 1 & 0 & 1 \\
C & 1 & 1 & 1 & 1 & 0 \\
D & 0 & 0 & 1 & 1 & 1 \\
E & 0 & 1 & 0 & 1 & 1 \\
\end{array}
$$

C.
$$
\begin{array}{c|ccccc}
 & A & B & C & D & E \\
\hline
A & 0 & 1 & 1 & 1 & 1 \\
B & 1 & 0 & 0 & 0 & 0 \\
C & 1 & 0 & 0 & 1 & 1 \\
D & 1 & 0 & 1 & 0 & 1 \\
E & 1 & 0 & 1 & 1 & 0 \\
\end{array}
$$

D.
$$
\begin{array}{c|ccccc}
 & A & B & C & D & E \\
\hline
A & 0 & 1 & 1 & 0 & 0 \\
B & 1 & 0 & 1 & 0 & 1 \\
C & 1 & 1 & 0 & 1 & 0 \\
D & 0 & 0 & 1 & 0 & 1 \\
E & 0 & 1 & 0 & 1 & 0 \\
\end{array}
$$

MM3A7b

45. What would the inverse matrix solution look like to the following system of equations?
$$
\begin{cases}
4x + 2y = 6 \\
10x + 4y = 6
\end{cases}
$$

A. $\begin{bmatrix} -4 & 2 \\ 10 & -4 \end{bmatrix}^{-1} \begin{bmatrix} 6 \\ 6 \end{bmatrix} = \begin{bmatrix} -3 \\ 9 \end{bmatrix}$

B. $\begin{bmatrix} -4 & 2 \\ 10 & -4 \end{bmatrix}^{-1} \begin{bmatrix} 6 \\ 6 \end{bmatrix} = \begin{bmatrix} 9 \\ 21 \end{bmatrix}$

C. $\begin{bmatrix} 4 & 2 \\ 10 & 4 \end{bmatrix}^{-1} \begin{bmatrix} 6 \\ 6 \end{bmatrix} = \begin{bmatrix} 9 \\ 21 \end{bmatrix}$

D. $\begin{bmatrix} 4 & 2 \\ 10 & 4 \end{bmatrix}^{-1} \begin{bmatrix} 6 \\ 6 \end{bmatrix} = \begin{bmatrix} -3 \\ 9 \end{bmatrix}$

MM3A5a

46. What system of inequalities defines the shaded areas below?

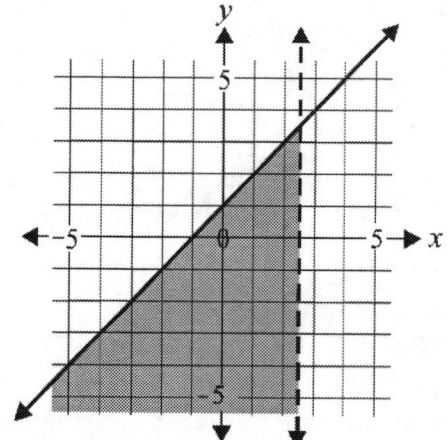

A. $y \leq x + 1$
$x < 2.5$

B. $y \geq x + 1$
$x > 2.5$

C. $y \leq x + 1$
$x > 2.5$

D. $y > x + 1$
$x < 2.5y$

MM3A6a

47. Two lines are shown on the grid. One line passes through the origin and the other passes through $(-1, -1)$ with a y-intercept of 2. Which pair of equations below the grid identifies these lines?

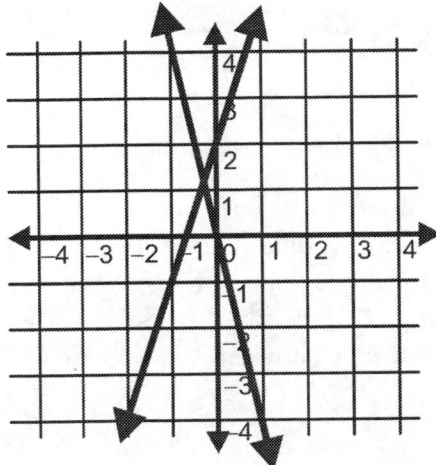

A. $y = \frac{1}{4}x$ and $y = \frac{1}{3}x + 2$

B. $x - 2y = 6$ and $4x + y = 4$

C. $y = 4x$ and $y = \frac{1}{3}x$

D. $y = 3x + 2$ and $y = -4x$

MM3A5c

48. $\begin{bmatrix} -13 & 2 \\ 5 & 8 \end{bmatrix} \times \begin{bmatrix} 3 & 6 \\ -7 & 11 \end{bmatrix} = ?$

A. $\begin{bmatrix} -53 & -41 \\ -56 & 118 \end{bmatrix}$

B. $\begin{bmatrix} -53 & -56 \\ -41 & 118 \end{bmatrix}$

C. $\begin{bmatrix} -41 & -53 \\ 118 & -56 \end{bmatrix}$

D. $\begin{bmatrix} -56 & -53 \\ 118 & -41 \end{bmatrix}$

MM3A4a

49. $-1\begin{bmatrix} 1 & -6 \\ 2 & -3 \end{bmatrix} + 3\begin{bmatrix} 3 & -1 \\ 0 & -3 \end{bmatrix} =$

A. $\begin{bmatrix} 8 & 3 \\ -2 & -6 \end{bmatrix}$

B. $\begin{bmatrix} 11 & 6 \\ 1 & 9 \end{bmatrix}$

C. $\begin{bmatrix} 4 & -7 \\ 2 & -6 \end{bmatrix}$

D. $\begin{bmatrix} 10 & -9 \\ 2 & -12 \end{bmatrix}$

MM3A4a

50. A student takes a true-false quiz consisting of 4 questions. The student is clueless and must guess at the answers. What is the probability that he guesses at most 2 correctly?

A. 68.75%
B. 50%
C. 22.25%
D. cannot be determined

MM3D1

51. There are 25 questions in a test that contain four multiple choices. What is the probability that without reading them you guessed 12 of the questions correctly?

A. 0
B. 0.0074
C. 0.0074%
D. 0.74

MM3D1

52. Given that a binomial distribution has a probability of $p = 0.23$ and $n = 12$, what is the most likely number of successes that will occur?

A. 3
B. 1
C. 6
D. 2

MM3D1

53. Construct a binomial distribution with your graphing calculator with $p = 0.76$ and $n = 22$. What is the most likely number of successes that will occur?

   A. 15

   B. 16

   C. 17

   D. 18

   MM3D1

54. For a normal distribution that has a mean of $\overline{x}$ and a standard deviation of σ, what is the probability of $P\left(\overline{x} \leq \overline{x} + 2\sigma\right)$?

   A. 0.977

   B. 0.9785

   C. 0.8410

   D. 0.997

   MM3D2a

55. What is the interval about the mean of a normally distributed data set that will include 81.8% of the data that has a mean of 56 and a standard deviation of 3?

   A. from 53 to 59

   B. from 50 to 59

   C. from 47 to 59

   D. from 56 to 59

   MM3D2a

56. What is the interval about the mean of a normally distributed data set that will include 47.7% of the data that has a mean of 61 and a standard deviation of 7?

   A. from 61 to 75

   B. from 54 to 68

   C. from 61 to 68

   D. from 54 to 61

   MM3D2a

57. What is the interval about the mean of a normally distributed data set that will include 83.95% of the data that has a mean of 96 and a standard deviation of 12?

   A. from 60 to 108

   B. from 72 to 108

   C. from 96 to 132

   D. from 72 to 120

   MM3D2a

58. If a normal distribution has a mean of 26 and a standard deviation of 1.3, what is the probability of randomly selecting a data point and it falls between 22.1 and 23.4?

   A. 0.0230

   B. 0.1575

   C. 0.0215

   D. 0.1360

   MM3D2b

59. What is the probability of selecting a random data point within a normal distribution that has a mean of 314 and a standard deviation of 5.6 and it falls between 308.4 and 325.2?

   A. 0.6820

   B. 0.3410

   C. 0.8180

   D. 0.4770

   MM3D2b

60. Given that a normal distribution has a mean of 35 and a standard deviation of 0.9, what is the probability that a randomly selected value from the distribution is greater than 36.8?

   A. 0.0230

   B. 0.9770

   C. 0.0215

   D. 0.8410

   MM3D2b

61. A gym records how long their members work out each day. There are $1,200$ members and their workout times are normally distributed with a mean of 45 minutes and a standard deviation of 4.2 minutes. How many people would you expect to work out between 36.6 minutes and 49.2 minutes?

    A. 818
    B. 982
    C. 600
    D. 572

    MM3D2c

62. What is the probability of selecting a random value from a normal distribution and it is at most 17 when the mean is 15.75 and the standard deviation is 1.25?

    A. 0.8410
    B. 0.1587
    C. 0.1590
    D. cannot be determined

    MM3D2b

63. Out of a sample of 600 males, their heart rate is normally distributed with a mean of 58 beats per minute and a standard deviation of 3 beats per minute. How many of those males would you expect to have heart rates that fall between 55 beats per minute and 61 beats per minute?

    A. 205
    B. 572
    C. 286
    D. 409

    MM3D2c

64. From the information in question 63, how many males would you expect to have a heart rate that falls between 61 beats per minute and 64 beats per minute?

    A. 164
    B. 205
    C. 82
    D. 95

    MM3D2c

65. What is the probability of selecting a value that is less than 7.5 out of a normal distribution that has a mean of 15 and a standard deviation of 2.5?

    A. 0.0215
    B. 0.0230
    C. 0.0015
    D. cannot be determined

    MM3D2b

66. The past $2,000$ births at a hospital had baby weights that are normally distributed with a mean of 7 pounds 4 ounces with a standard deviation of 1 pound 2 ounces. How many of those babies would you expect to have weighed between 8 pounds 6 ounces and 10 pounds 10 ounces?

    A. 272
    B. 954
    C. 315
    D. 43

    MM3D2c

67. The mileage when people get their oil changed in their car is normally distributed with a mean of $3,000$ miles and a standard deviation of 200 miles. From the past 300 oil changes at a particular oil change company, how many customers do you expect to have driven at least $2,800$ miles since their last oil change?

    A. 48
    B. 252
    C. 150
    D. cannot be determined

    MM3D2c

68. What kind of sample is taken when there is a selection rule such as people are numbered 1 through 3 and the sample consists of only people who are number 2?

    A. Random Sample
    B. Systematic Sample
    C. Self-selected Sample
    D. Convenience Sample

    MM3D3

69. According to a recent survey, 23% of homeowners hand wash their dishes rather than using a dishwasher. If you were to survey 1,350 homeowners, what is the approximate probability that up to 295 of the homeowners hand wash their dishes?

    A. 0.3410
    B. 0.0015
    C. 0.1590
    D. 0.0215

MM3D3

70. In a survey of 3,500 people, 62% said that they run at least once a week. What is the margin of error for this survey?

    A. $\pm 0.0029\%$
    B. $\pm 1.69\%$
    C. $\pm 1.61\%$
    D. $\pm 12.70\%$

MM3D3

71. A double blind study is where the people who are placed in the experimental group and control group are randomly placed and the people conducting the survey have no idea who is in the experimental group or control group. What kind of study is a double blind study?

    A. Observational Study
    B. Experimental Study
    C. Random Study
    D. Conclusive Study

MM3D3

72. Solve the equation $AX = B$ for X.

$$A = \begin{bmatrix} 2 & -1 \\ 3 & 0 \end{bmatrix} \text{ and } B = \begin{bmatrix} -10 & -1 \\ -15 & 12 \end{bmatrix}$$

    A. $X = \begin{bmatrix} -5 & 4 \\ 0 & 9 \end{bmatrix}$

    B. $X = \begin{bmatrix} -5 & 1 \\ -5 & 0 \end{bmatrix}$

    C. $X = \begin{bmatrix} -5 & -14 \\ -30 & -3 \end{bmatrix}$

    D. $X = \begin{bmatrix} -23 & 10 \\ 6 & 15 \end{bmatrix}$

MM3A5b

Evaluation Chart for the Diagnostic Mathematics Test

Directions: On the following chart, circle the question numbers that you answered incorrectly. Then turn to the appropriate topics (listed by chapters), read the explanations, and complete the exercises. Review the other chapters as needed. Finally, complete the *Mastering the Georgia Mathematics III Course* Practice Test to further review.

		Questions	Pages
Chapter 1:	Matrices	44, 48, 49, 72	14–35
Chapter 2:	Systems of Equations and Inequalities	43, 45, 46, 47	36–51
Chapter 3:	Circles	1, 2, 3, 4, 5, 6, 7, 8, 9, 10, 24, 25	52–62
Chapter 4:	Conics	15, 16, 17, 18, 19, 20, 21	63–74
Chapter 5:	Exponential and Logarithmic Functions	26, 27, 28, 29, 30	75–99
Chapter 6:	Solving Polynomial, Exponential, and Logarithmic Equations	31, 32, 33, 34	100–140
Chapter 7:	Solving Polynomial, Exponential, and Logarithmic Inequalities	35, 36, 37, 38	141–170
Chapter 8:	Graphing Polynomial Functions	39, 40, 41, 42	171–197
Chapter 9:	Three-Dimensional Space	11, 12, 13, 14, 22, 23	198–207
Chapter 10:	Probability	50, 51, 52, 53, 54, 55, 56, 57, 58, 59, 60, 61, 62, 63, 64, 65, 66, 67	208–221
Chapter 11:	Data Interpretation	68, 69, 70, 71	222–232

Chapter 1
Matrices

This chapter covers the following Georgia Performance Standards:

MM3A	Algebra	MM3A4a, MM3A4b, MM3A4c
		MM3A5b
		MM3A7a, MM3A7b

A **matrix** (plural: **matrices**) is a rectangular or square ordered array of numbers, and each number in a matrix is called an **element**. The matrix shown below contains six elements: 3, −1, 2, 4, 0, and 1. It is arranged in two rows and three columns and, therefore, is referred to as a 2×3 matrix. When describing a matrix, always give the number of rows and then the number of columns.

$$\textbf{2 Rows} \begin{bmatrix} 3 & -1 & 2 \\ 4 & 0 & 1 \end{bmatrix}$$

3 Columns

Since their invention in 1858, matrices have played a role in such fields as economics, engineering, and quantum mechanics. This chapter will cover the use of basic operations on matrices.

1.1 Addition of Matrices

In order to add matrices, they must be of the same size; they need to have the same number of rows and columns. A 2×3 matrix can only be added to another 2×3 matrix.

To add two matrices of the same size, add the corresponding elements of the two matrices. The resulting matrix is the same size as each of the two matrices that were added together.

Example 1:
$$\begin{bmatrix} 7 & -2 \\ -1 & 4 \end{bmatrix} + \begin{bmatrix} -6 & 2 \\ 4 & 0 \end{bmatrix} = \begin{bmatrix} 7+(-6) & (-2)+2 \\ (-1)+4 & 4+0 \end{bmatrix} = \begin{bmatrix} 1 & 0 \\ 3 & 4 \end{bmatrix}$$

Note that the resulting matrix is a 2×2 matrix, as are the matrices added together.

Example 2:
$$\begin{bmatrix} 7 & -2 \\ -1 & 4 \\ 5 & -3 \end{bmatrix} + \begin{bmatrix} -6 & 2 & 2 \\ 4 & 0 & 0 \end{bmatrix} =$$

The matrices to be added are not of the same size. The first matrix is a 3×2 matrix, and the second matrix is 2×3. Therefore, these two matrices cannot be added. It is not possible to add these matrices.

Note: Any number of matrices of the same size can be added together.

Add the matrices together when possible. When the matrices cannot be added, write NP.

1. $\begin{bmatrix} 8 & 4 \\ 5 & -3 \end{bmatrix} + \begin{bmatrix} 0 & -7 \\ -4 & -9 \end{bmatrix}$

2. $\begin{bmatrix} 6 \\ -4 \\ 5 \end{bmatrix} + \begin{bmatrix} 2 \\ -9 \\ -8 \end{bmatrix} + \begin{bmatrix} 9 \\ -1 \\ -3 \end{bmatrix} + \begin{bmatrix} 5 \\ -1 \\ -2 \end{bmatrix}$

3. $\begin{bmatrix} 3 & -1 & 2 \\ 4 & 0 & 1 \end{bmatrix} + \begin{bmatrix} 6 & 0 & 5 \\ 4 & 0 & 1 \end{bmatrix}$

4. $\begin{bmatrix} -6 \\ -2 \end{bmatrix} + \begin{bmatrix} -4 & -1 \end{bmatrix}$

5. $\begin{bmatrix} -5 & -2 \\ 9 & -7 \\ 3 & 6 \end{bmatrix} + \begin{bmatrix} -1 & -2 \\ -8 & 7 \\ 9 & -4 \end{bmatrix}$

6. $\begin{bmatrix} 8 & -1 & -6 & 3 \\ 0 & -7 & -5 & -4 \end{bmatrix} + \begin{bmatrix} -8 & 1 & 5 & -2 \\ 1 & 6 & 6 & 5 \end{bmatrix}$

1.2 Multiplication of a Matrix by a Constant

A matrix can be multiplied by a constant. The constant is multiplied by each element in the matrix. The resulting matrix is the same size as the matrix being multiplied.

Example 3: $4 \begin{bmatrix} 1 & 0 \\ 3 & 4 \end{bmatrix} =$

Step 1: Multiply every number in the matrix by 4.

$$\begin{bmatrix} (4 \times 1) & (4 \times 0) \\ (4 \times 3) & (4 \times 4) \end{bmatrix} = \begin{bmatrix} 4 & 0 \\ 12 & 16 \end{bmatrix}$$

Step 2: The solution of $4 \begin{bmatrix} 1 & 0 \\ 3 & 4 \end{bmatrix}$ is $\begin{bmatrix} 4 & 0 \\ 12 & 16 \end{bmatrix}$.

Example 4: $-3 \begin{bmatrix} -8 & 1 & 5 & -2 \\ 1 & 6 & 6 & 5 \end{bmatrix} =$

Step 1: Multiply every number in the matrix by -3.

$$\begin{bmatrix} (-3 \times -8) & (-3 \times 1) & (-3 \times 5) & (-3 \times -2) \\ (-3 \times 1) & (-3 \times 6) & (-3 \times 6) & (-3 \times 5) \end{bmatrix} = \begin{bmatrix} 24 & -3 & -15 & 6 \\ -3 & -18 & -18 & -15 \end{bmatrix}$$

Step 2: The solution of $-3 \begin{bmatrix} -8 & 1 & 5 & -2 \\ 1 & 6 & 6 & 5 \end{bmatrix}$ is $\begin{bmatrix} 24 & -3 & -15 & 6 \\ -3 & -18 & -18 & -15 \end{bmatrix}$

Multiply each of the following.

1. $-2 \begin{bmatrix} -8 & 1 \\ 6 & 7 \\ -5 & -4 \end{bmatrix}$

2. $5 \begin{bmatrix} 3 \\ 0 \\ -1 \end{bmatrix}$

3. $\dfrac{1}{2} \begin{bmatrix} 2 & -4 & -9 & 7 \\ -1 & 6 & 3 & -8 \end{bmatrix}$

4. $-\dfrac{1}{4} \begin{bmatrix} -1 & -6 & 4 & -2 \\ 8 & 0 & 3 & -8 \\ 4 & 6 & -3 & 12 \end{bmatrix}$

5. $-9 \begin{bmatrix} -10 & -1 & \dfrac{1}{3} & 7 & -\dfrac{3}{4} \end{bmatrix}$

6. $6 \begin{bmatrix} -4 & -5 \\ 1 & 0 \end{bmatrix}$

1.3   Subtraction of Matrices

Subtraction of matrices is similar to addition of matrices in that the matrices to be subtracted must be the same size. Suppose x and y represent two different matrices of the same size. $x - y$ can also be written $x + (-1)\,y$. Therefore, subtraction of matrices involves two steps: multiplying the second matrix by -1 and then adding it to the first matrix.

Example 5: $\begin{bmatrix} 4 & -6 \\ 9 & -5 \end{bmatrix} - \begin{bmatrix} 3 & -1 \\ -4 & 7 \end{bmatrix}$ can also be written as $\begin{bmatrix} 4 & -6 \\ 9 & -5 \end{bmatrix} + (-1) \begin{bmatrix} 3 & -1 \\ -4 & 7 \end{bmatrix}$

Step 1:   Multiply the second matrix by -1.

$$(-1) \begin{bmatrix} 3 & -1 \\ -4 & 7 \end{bmatrix} = \begin{bmatrix} -3 & 1 \\ 4 & -7 \end{bmatrix}$$

Step 2:   Add the first matrix and the product from step 1.

$$\begin{bmatrix} 4 & -6 \\ 9 & -5 \end{bmatrix} + \begin{bmatrix} -3 & 1 \\ 4 & -7 \end{bmatrix} = \begin{bmatrix} 1 & -5 \\ 13 & -12 \end{bmatrix}$$

Subtract the matrices when possible. When the matrices cannot be subtracted, write NP.

1. $\begin{bmatrix} 1 & 2 \\ 6 & 0 \\ -1 & 4 \end{bmatrix} - \begin{bmatrix} 3 & 2 \\ 1 & -3 \\ 5 & 1 \end{bmatrix}$

4. $\begin{bmatrix} 2 & 3 \\ 1 & 5 \end{bmatrix} - \begin{bmatrix} 1 & 3 \\ -2 & 1 \end{bmatrix}$

2. $\begin{bmatrix} 1 & 2 & 3 \\ 4 & 5 & 6 \end{bmatrix} - \begin{bmatrix} 3 & 2 & 1 \\ 5 & 4 & 6 \end{bmatrix}$

5. $\begin{bmatrix} 8 & -1 & -6 & 3 \\ 0 & -7 & -5 & -4 \end{bmatrix} - \begin{bmatrix} -8 & 1 & 5 & -2 \\ 1 & 6 & 6 & 5 \end{bmatrix}$

3. $\begin{bmatrix} 6 & 2 & 2 \\ -4 & -3 & -9 \\ 5 & 1 & -8 \end{bmatrix} - \begin{bmatrix} 6 & 9 & -4 \\ -4 & 0 & -8 \\ 2 & 3 & -7 \end{bmatrix}$

6. $\begin{bmatrix} 1 & 2 & 5 & 5 \\ 7 & 9 & 1 & -8 \\ -3 & 2 & 5 & -7 \end{bmatrix} - \begin{bmatrix} 1 & 3 & 1 \\ -1 & 0 & -8 \\ 0 & 1 & -7 \end{bmatrix}$

Perform the proper operation(s) on each set of matrices.

7. $\begin{bmatrix} 4 & -1 & 2 & 7 \\ -3 & 1 & -5 & -4 \end{bmatrix} + \dfrac{1}{2} \begin{bmatrix} 6 & 8 & 10 & -2 \\ -4 & 2 & 0 & -6 \end{bmatrix}$

8. $4 \begin{bmatrix} -1 & 2 & -\dfrac{1}{2} & 0 & -\dfrac{3}{4} \end{bmatrix} - \begin{bmatrix} -1 & 0 & -5 & 6 & -3 \end{bmatrix}$

1.4   More Practice

Example 6:   Given: $A = \begin{bmatrix} 1 & 4 & 7 \\ 2 & 5 & 8 \end{bmatrix}$ $B = \begin{bmatrix} 1 & 2 & 3 \\ 4 & 5 & 6 \end{bmatrix}$

Find $2A + B$.

Step 1:   Plug matrix A and B into the given equation.

$$2A + B = 2 \begin{bmatrix} 1 & 4 & 7 \\ 2 & 5 & 8 \end{bmatrix} + \begin{bmatrix} 1 & 2 & 3 \\ 4 & 5 & 6 \end{bmatrix}$$

Step 2:   Multiply every number in matrix A by 2.

$$2 \begin{bmatrix} 1 & 4 & 7 \\ 2 & 5 & 8 \end{bmatrix} = \begin{bmatrix} (2 \times 1) & (2 \times 4) & (2 \times 7) \\ (2 \times 2) & (2 \times 5) & (2 \times 8) \end{bmatrix} = \begin{bmatrix} 2 & 8 & 14 \\ 4 & 10 & 16 \end{bmatrix}$$

Step 3:   Add the corresponding elements of the two matrices.

$$\begin{bmatrix} 2 & 8 & 14 \\ 4 & 10 & 16 \end{bmatrix} + \begin{bmatrix} 1 & 2 & 3 \\ 4 & 5 & 6 \end{bmatrix} = \begin{bmatrix} 2+1 & 8+2 & 14+3 \\ 4+4 & 10+5 & 16+6 \end{bmatrix} = \begin{bmatrix} 3 & 10 & 17 \\ 8 & 15 & 22 \end{bmatrix}$$

Solve the following problems.

1. $\begin{bmatrix} 7 & 15 & 11 \\ 1 & 2 & 8 \\ 9 & 3 & 12 \end{bmatrix} + 2 \begin{bmatrix} 3 & 2 & 1 \\ 5 & 6 & 7 \\ 8 & 9 & 0 \end{bmatrix}$

3. $4 \begin{bmatrix} 3 & 0 & 2 \\ 9 & 3 & 8 \\ 7 & 5 & 3 \end{bmatrix} + \begin{bmatrix} -2 & 0 & 12 \\ 4 & -5 & 6 \\ 1 & 8 & 5 \end{bmatrix}$

2. $3 \begin{bmatrix} 2 & 4 \\ 6 & 8 \end{bmatrix} - 3 \begin{bmatrix} 9 & 7 \\ 5 & 3 \end{bmatrix}$

4. $6 \begin{bmatrix} 1 & 7 \\ 5 & 10 \end{bmatrix} - \begin{bmatrix} 6 & 6 \\ 5 & -3 \\ 1 & 0 \end{bmatrix}$

Use the following matrices to solve problems 5–10.

$A = \begin{bmatrix} -2 & 6 \\ 5 & -4 \\ 9 & -7 \end{bmatrix}$ $B = \begin{bmatrix} 5 & -5 & 2 \\ -1 & 8 & 6 \\ 3 & 10 & 8 \end{bmatrix}$ $C = \begin{bmatrix} -9 & -11 \\ 0 & 5 \\ 3 & -4 \end{bmatrix}$

$D = \begin{bmatrix} 0 & -6 & 11 \\ 1 & 5 & 5 \\ 5 & 8 & -3 \end{bmatrix}$ $E = \begin{bmatrix} 8 & -2 \\ 6 & 0 \\ -7 & 6 \end{bmatrix}$ $F = \begin{bmatrix} 10 & 16 & 0 \\ 5 & -8 & -2 \\ 6 & -7 & 5 \end{bmatrix}$

5. Find $2A + 3E$.

7. Find $3C + 2A - E$.

9. Find $-C + 3A$.

6. Find $2F - B$.

8. Find $4D - 2B$.

10. Find $2D - 2B + 2F$.

1.5   Multiplying Matrices

Example 7:   Multiply: $\begin{bmatrix} 9 & 1 & 6 \\ 8 & 2 & 7 \\ 7 & -3 & 8 \\ 6 & 4 & 9 \\ 5 & 5 & 0 \end{bmatrix} \times \begin{bmatrix} 4 & 1 \\ 6 & 0 \\ 8 & -1 \end{bmatrix} =$

Step 1:   Determine the size of the resulting matrix.
When multiplying matrices, the number of columns in the first matrix must equal the number of rows in the second matrix. The resulting matrix will consist of the number of rows in the first matrix by the number of columns in the second matrix.
The first matrix is a 5×3 matrix and the second matrix is a 3×2 matrix, so the resulting matrix will be a 5×2 matrix.

$$\begin{bmatrix} 9 & 1 & 6 \\ 8 & 2 & 7 \\ 7 & -3 & 8 \\ 6 & 4 & 9 \\ 5 & 5 & 0 \end{bmatrix} \times \begin{bmatrix} 4 & 1 \\ 6 & 0 \\ 8 & -1 \end{bmatrix} = \begin{bmatrix} \_ & \_ \\ \_ & \_ \\ \_ & \_ \\ \_ & \_ \\ \_ & \_ \end{bmatrix}$$

Step 2:   Multiply the first row in the first matrix by the first column in second matrix.

$$\begin{bmatrix} 9 & 1 & 6 \end{bmatrix} \times \begin{bmatrix} 4 \\ 6 \\ 8 \end{bmatrix} = (9 \times 4) + (1 \times 6) + (6 \times 8) = 36 + 6 + 48 = 90$$

The number, 90, will go in the first row and first column of the resulting matrix.

$$\begin{bmatrix} 90 & \_ \\ \_ & \_ \\ \_ & \_ \\ \_ & \_ \end{bmatrix}$$

Step 3:   Next multiply the second row in the first matrix by the first column in the second matrix.

$$\begin{bmatrix} 8 & 2 & 7 \end{bmatrix} \times \begin{bmatrix} 4 \\ 6 \\ 8 \end{bmatrix} = (8 \times 4) + (2 \times 6) + (7 \times 8) = 100$$

The number, 100, will go in the second row and first column of the resulting matrix.

$$\begin{bmatrix} 90 & \_ \\ 100 & \_ \\ \_ & \_ \\ \_ & \_ \end{bmatrix}$$

Step 4:   Continue multiplying in the same manner for rows three through five of the first matrix.

Row 3 by column 1: $\begin{bmatrix} 7 & -3 & 8 \end{bmatrix} \times \begin{bmatrix} 4 \\ 6 \\ 8 \end{bmatrix} = (7 \times 4) + (-3 \times 6) + (8 \times 8) = 74$

Row 4 by column 1: $\begin{bmatrix} 6 & 4 & 9 \end{bmatrix} \times \begin{bmatrix} 4 \\ 6 \\ 8 \end{bmatrix} = (6 \times 4) + (4 \times 6) + (9 \times 8) = 120$

Row 5 by column 1: $\begin{bmatrix} 5 & 5 & 0 \end{bmatrix} \times \begin{bmatrix} 4 \\ 6 \\ 8 \end{bmatrix} = (5 \times 4) + (5 \times 6) + (0 \times 8) = 50$

Plug the values into the appropriate spaces: $\begin{bmatrix} 90 & \text{—} \\ 100 & \text{—} \\ 74 & \text{—} \\ 120 & \text{—} \\ 50 & \text{—} \end{bmatrix}$

Step 5:   Now multiply all of the rows in the first matrix by the second column in the second matrix.

Row 1 by column 2: $\begin{bmatrix} 9 & 1 & 6 \end{bmatrix} \times \begin{bmatrix} 1 \\ 0 \\ -1 \end{bmatrix} = (9 \times 1) + (1 \times 0) + (6 \times -1) = 3$

Row 2 by column 2: $\begin{bmatrix} 8 & 2 & 7 \end{bmatrix} \times \begin{bmatrix} 1 \\ 0 \\ -1 \end{bmatrix} = (8 \times 1) + (2 \times 0) + (7 \times -1) = 1$

Row 3 by column 2: $\begin{bmatrix} 7 & -3 & 8 \end{bmatrix} \times \begin{bmatrix} 1 \\ 0 \\ -1 \end{bmatrix} = (7 \times 1) + (-3 \times 0) + (8 \times -1) = -1$

Row 4 by column 2: $\begin{bmatrix} 6 & 4 & 9 \end{bmatrix} \times \begin{bmatrix} 1 \\ 0 \\ -1 \end{bmatrix} = (6 \times 1) + (4 \times 0) + (9 \times -1) = -3$

Row 5 by column 2: $\begin{bmatrix} 5 & 5 & 0 \end{bmatrix} \times \begin{bmatrix} 1 \\ 0 \\ -1 \end{bmatrix} = (5 \times 1) + (5 \times 0) + (0 \times -1) = 5$

Step 6:   Plug all of the values found in step 4 in column 2 of the resulting matrix to find the answer.

$$\begin{bmatrix} 9 & 1 & 6 \\ 8 & 2 & 7 \\ 7 & -3 & 8 \\ 6 & 4 & 9 \\ 5 & 5 & 0 \end{bmatrix} \times \begin{bmatrix} 4 & 1 \\ 6 & 0 \\ 8 & -1 \end{bmatrix} = \begin{bmatrix} 90 & 3 \\ 100 & 1 \\ 74 & -1 \\ 120 & -3 \\ 50 & 5 \end{bmatrix}$$

Example 8:   Multiply: $\begin{bmatrix} -1 & 0 \\ 0 & -1 \end{bmatrix} \times \begin{bmatrix} -4 & -3 & 3 & -7 \\ 2 & -6 & -5 & 8 \end{bmatrix} =$

Step 1:   Determine the size of the resulting matrix.
The first matrix is a 2×2 matrix and the second matrix is a 2×4 matrix, so the resulting matrix will be a 2×4 matrix.

$$\begin{bmatrix} -1 & 0 \\ 0 & -1 \end{bmatrix} \times \begin{bmatrix} -4 & -3 & 3 & -7 \\ 2 & -6 & -5 & 8 \end{bmatrix} = \begin{bmatrix} - & - & - & - \\ - & - & - & - \end{bmatrix}$$

Step 2:   Now multiply all of the rows in the first matrix by the first column in the second matrix.

Row 1 by column 1: $\begin{bmatrix} -1 & 0 \end{bmatrix} \times \begin{bmatrix} -4 \\ 2 \end{bmatrix} = (-1 \times -4) + (0 \times 2) = 4$

Row 2 by column 1: $\begin{bmatrix} 0 & -1 \end{bmatrix} \times \begin{bmatrix} -4 \\ 2 \end{bmatrix} = (0 \times -4) + (-1 \times 2) = -2$

Plug the values into the appropriate spaces: $\begin{bmatrix} 4 & - & - & - \\ -2 & - & - & - \end{bmatrix}$

Step 3:   Now multiply all of the rows in the first matrix by the first column in the second matrix.

Row 1 by column 2: $\begin{bmatrix} -1 & 0 \end{bmatrix} \times \begin{bmatrix} -3 \\ -6 \end{bmatrix} = (-1 \times -3) + (0 \times -6) = 3$

Row 2 by column 2: $\begin{bmatrix} 0 & -1 \end{bmatrix} \times \begin{bmatrix} -3 \\ -6 \end{bmatrix} = (0 \times -3) + (-1 \times -6) = 6$

Plug the values into the appropriate spaces: $\begin{bmatrix} 4 & 3 & - & - \\ -2 & 6 & - & - \end{bmatrix}$

Step 4:   Now multiply all of the rows in the first matrix by the first column in the second matrix.

Row 1 by column 3: $\begin{bmatrix} -1 & 0 \end{bmatrix} \times \begin{bmatrix} 3 \\ -5 \end{bmatrix} = (-1 \times 3) + (0 \times -5) = -3$

Row 2 by column 3: $\begin{bmatrix} 0 & -1 \end{bmatrix} \times \begin{bmatrix} 3 \\ -5 \end{bmatrix} = (0 \times 3) + (-1 \times -5) = 5$

Plug the values into the appropriate spaces: $\begin{bmatrix} 4 & 3 & -3 & - \\ -2 & 6 & 5 & - \end{bmatrix}$

Step 5: Now multiply all of the rows in the first matrix by the first column in the second matrix.

Row 1 by column 4: $\begin{bmatrix} -1 & 0 \end{bmatrix} \times \begin{bmatrix} -7 \\ 8 \end{bmatrix} = (-1 \times -7) + (0 \times 8) = 7$

Row 2 by column 4: $\begin{bmatrix} 0 & -1 \end{bmatrix} \times \begin{bmatrix} -7 \\ 8 \end{bmatrix} = (0 \times -7) + (-1 \times 8) = -8$

Plug the values into the appropriate spaces: $\begin{bmatrix} 4 & 3 & -3 & 7 \\ -2 & 6 & 5 & -8 \end{bmatrix}$

Step 6: $\begin{bmatrix} -1 & 0 \\ 0 & -1 \end{bmatrix} \times \begin{bmatrix} -4 & -3 & 3 & -7 \\ 2 & -6 & -5 & 8 \end{bmatrix} = \begin{bmatrix} 4 & 3 & -3 & 7 \\ -2 & 6 & 5 & -8 \end{bmatrix}$

Multiply the following matrices.

1. $\begin{bmatrix} 4 & 2 \\ 1 & 3 \end{bmatrix} \times \begin{bmatrix} 2 & -6 & 8 \\ -4 & 3 & 1 \end{bmatrix} =$

5. $\begin{bmatrix} 0 & -1 \\ -1 & 0 \end{bmatrix} \times \begin{bmatrix} -4 & 2 & 2 & -4 \\ 5 & 3 & -3 & 5 \end{bmatrix} =$

2. $\begin{bmatrix} 1 & 0 \\ 0 & 1 \end{bmatrix} \times \begin{bmatrix} -4 & 8 & 7 \\ 2 & 3 & -5 \end{bmatrix} =$

6. $\begin{bmatrix} 3 & 6 & 9 & -4 \\ 8 & 1 & -5 & 2 \end{bmatrix} \times \begin{bmatrix} -8 \\ -4 \\ -1 \\ 2 \end{bmatrix} =$

3. $\begin{bmatrix} -5 & 4 & 3 \\ 2 & -1 & 6 \end{bmatrix} \times \begin{bmatrix} 1 \\ 2 \\ 3 \end{bmatrix} =$

7. $\begin{bmatrix} 4 & 2 & 6 & 1 & 9 \\ 1 & 3 & 4 & 1 & 2 \end{bmatrix} \times \begin{bmatrix} 5 & 6 \\ 3 & 2 \\ 1 & 7 \\ 3 & 9 \\ 2 & 4 \end{bmatrix} =$

4. $\begin{bmatrix} 8 & 9 & 0 \\ -1 & -2 & 3 \end{bmatrix} \times \begin{bmatrix} 4 \\ 5 \\ 6 \end{bmatrix} =$

8. $\begin{bmatrix} 1 & -1 \\ -1 & 1 \end{bmatrix} \times \begin{bmatrix} 4 & 3 & 1 \\ 2 & 1 & 2 \end{bmatrix} =$

1.6   Determinants of Matrices

The **determinant** of the 2×2 matrix $\begin{bmatrix} a & b \\ c & d \end{bmatrix}$ is $ad - bc$, while the determinant of the 3×3

matrix $\begin{bmatrix} a_1 & b_1 & c_1 \\ a_2 & b_2 & c_2 \\ a_3 & b_3 & c_3 \end{bmatrix}$ is $a_1b_2c_3 + b_1c_2a_3 + c_1a_2b_3 - a_3b_2c_1 - b_3c_2a_1 - c_3a_2b_1$. Determinants

of matrices are used for many purposes, including the calculation of inverses. If $A = \begin{bmatrix} a & b \\ c & d \end{bmatrix}$,

and if the determinant of A (written $\det A$) does not equal 0, the inverse of A (written A^{-1}) is

$\dfrac{1}{\det A} \begin{bmatrix} d & -b \\ -c & a \end{bmatrix}$, or $\dfrac{1}{ad - bc} \begin{bmatrix} d & -b \\ -c & a \end{bmatrix}$.

Note: It's not necessary to memorize the formulas for calculating the determinants of matrices, as they will be given.

Example 9:   Find the inverse of the matrix $A = \begin{bmatrix} -5 & 2 \\ 3 & -4 \end{bmatrix}$.

Step 1:   First, the determinant of A is calculated with the formula $\det A = ad - bc$ as follows: $(-5)(-4) - (2)(3) = 20 - 6 = 14$.

Step 2:   Next, the determinant of A and the values of d, $-b$, $-c$, and a are substituted into the formula for the inverse of matrix A:
$$A^{-1} = \frac{1}{\det A} \begin{bmatrix} d & -b \\ -c & a \end{bmatrix} = \frac{1}{14} \begin{bmatrix} -4 & -2 \\ -3 & -5 \end{bmatrix}$$

Step 3:   Finally, each element of the matrix $\begin{bmatrix} -4 & -2 \\ -3 & -5 \end{bmatrix}$ is multiplied by $\frac{1}{14}$, and the result is
$$\begin{bmatrix} -\frac{2}{7} & -\frac{1}{7} \\ -\frac{3}{14} & -\frac{5}{14} \end{bmatrix}.$$

Calculate the determinant of each of the following matrices.

1. $\begin{bmatrix} 3 & 8 \\ -1 & 2 \end{bmatrix}$

3. $\begin{bmatrix} 10 & 6 \\ 1 & 4 \end{bmatrix}$

5. $\begin{bmatrix} -9 & 1 \\ -8 & 2 \end{bmatrix}$

2. $\begin{bmatrix} 5 & -2 \\ -3 & 9 \end{bmatrix}$

4. $\begin{bmatrix} -4 & 7 \\ -5 & -7 \end{bmatrix}$

6. $\begin{bmatrix} -6 & -3 \\ -10 & 8 \end{bmatrix}$

Find the inverse of each of the following matrices.

7. $\begin{bmatrix} 4 & 3 \\ 2 & -1 \end{bmatrix}$

8. $\begin{bmatrix} 8 & -9 \\ 3 & -8 \end{bmatrix}$

9. $\begin{bmatrix} 11 & -1 \\ -2 & 3 \end{bmatrix}$

Example 10:   Calculate the determinant of the matrix $A = \begin{bmatrix} 7 & -9 & 1 \\ 4 & -3 & 2 \\ -5 & 6 & 8 \end{bmatrix}$.

Step 1:   To calculate the determinant of matrix A, use the formula
$\det A = a_1 b_2 c_3 + b_1 c_2 a_3 + c_1 a_2 b_3 - a_3 b_2 c_1 - b_3 c_2 a_1 - c_3 a_2 b_1$.

Step 2:   Substitute known values into the formula as shown.

$\det A = (7)(-3)(8) + (-9)(2)(-5) + (1)(4)(6) - (-5)(-3)(1) - (6)(2)(7) - (8)(4)(-9)$

Step 3:   Solve.

$\det A = -168 + 90 + 24 - 15 - 84 - (-288) = 135$.

Therefore, the determinant of matrix A is 135.

Calculate the determinant of each of the following matrices.

10. $\begin{bmatrix} 11 & 7 & 4 \\ -2 & -8 & 3 \\ 6 & 9 & 2 \end{bmatrix}$

13. $\begin{bmatrix} 6 & 3 & 7 \\ 7 & 2 & 6 \\ -8 & -3 & 5 \end{bmatrix}$

16. $\begin{bmatrix} 5 & -10 & 3 \\ 10 & 1 & -5 \\ -4 & 2 & 4 \end{bmatrix}$

11. $\begin{bmatrix} 3 & -5 & 9 \\ 2 & -6 & 8 \\ 1 & -7 & 7 \end{bmatrix}$

14. $\begin{bmatrix} 1 & 4 & 7 \\ -2 & -5 & -8 \\ -3 & -6 & -9 \end{bmatrix}$

17. $\begin{bmatrix} 9 & 6 & -3 \\ 3 & 7 & 2 \\ 1 & -12 & -2 \end{bmatrix}$

12. $\begin{bmatrix} 9 & 3 & 9 \\ 10 & 5 & 7 \\ -1 & -7 & -5 \end{bmatrix}$

15. $\begin{bmatrix} -1 & 2 & 11 \\ 5 & 10 & -2 \\ 12 & -3 & 3 \end{bmatrix}$

18. $\begin{bmatrix} 8 & -5 & 2 \\ -7 & 4 & -1 \\ 6 & -3 & 2 \end{bmatrix}$

19. The area of a triangle with vertices at the points (a, b), (c, d), and (e, f) is $\frac{1}{2}|\det A|$, where
$A = \begin{bmatrix} a & b & 1 \\ c & d & 1 \\ e & f & 1 \end{bmatrix}$.

What is the area of a triangle with vertices at the points $(3, 8)$, $(-5, 14)$ and $(6, 9)$?

1.7   Inverses of Matrices

As shown in the previous section, if $A = \begin{bmatrix} a & b \\ c & d \end{bmatrix}$ and the determinant of A does not equal 0,

the inverse of A, A^{-1}, is $\dfrac{1}{\det A} \begin{bmatrix} d & -b \\ -c & a \end{bmatrix}$, or $\dfrac{1}{ad-bc} \begin{bmatrix} d & -b \\ -c & a \end{bmatrix}$. For square matrices larger than 2×2, use a graphing calculator to find the inverse.

Example 11:   Find the inverse of the matrix $A = \begin{bmatrix} 1 & 0 & 2 \\ -2 & 1 & -4 \\ 1 & 2 & 0 \end{bmatrix}$.

Step 1:   First, put the 3×3 matrix into the MATRIX function of the calculator.

Step 2:   Next, calculate A^{-1} by pressing enter when in the matrix function, then the inverse key, which often looks like x^{-1}.

Step 3: $A^{-1} = \begin{bmatrix} -4 & -2 & 1 \\ 2 & 1 & 0 \\ 2.5 & 1 & -0.5 \end{bmatrix}$.

In mathematics, when a number is multiplied by its inverse, the product is 1. For example, $4 \times \frac{1}{4} = 1$. When a matrix is multiplied by its inverse, it does not equal 1. The product is the **identity matrix**. The identity matrix, notated as I, has 1's on the main diagonal and zeros everywhere else. The 2×2 identity matrix is $I = \begin{bmatrix} 1 & 0 \\ 0 & 1 \end{bmatrix}$.

Example 12:   Determine if matrices A and B are inverses of each other.

$$A = \begin{bmatrix} 10 & 3 \\ -3 & -1 \end{bmatrix} \text{ and } B = \begin{bmatrix} 1 & 3 \\ -3 & -10 \end{bmatrix}$$

Step 1:   The matrices A and B are inverses of each other if their product in both orders equals I.

Step 2:   Find AB.

$$AB = \begin{bmatrix} 10 & 3 \\ -3 & -1 \end{bmatrix} \begin{bmatrix} 1 & 3 \\ -3 & -10 \end{bmatrix} = \begin{bmatrix} 1 & 0 \\ 0 & 1 \end{bmatrix} = I$$

Step 3:   Find BA.

$$BA = \begin{bmatrix} 1 & 3 \\ -3 & -10 \end{bmatrix} \begin{bmatrix} 10 & 3 \\ -3 & -1 \end{bmatrix} = \begin{bmatrix} 1 & 0 \\ 0 & 1 \end{bmatrix} = I$$

Step 4:   Both AB and BA equal I, so A and B are inverses of each other.

Find the inverse of each matrix below using a graphing calculator. If no inverse exists for the matrix, write none. Give answers in fraction form.

1. $\begin{bmatrix} -1 & 1 & 0 \\ -1 & 0 & 1 \\ 6 & -2 & -3 \end{bmatrix}$

5. $\begin{bmatrix} 3 & -2 & 1 \\ 2 & 1 & 0 \\ 0 & 1 & -1 \end{bmatrix}$

9. $\begin{bmatrix} 11 & -7 & -2 \\ -17 & 11 & 3 \\ 2 & -1 & 0 \end{bmatrix}$

2. $\begin{bmatrix} -3 & -2 & 0 & -2 & -3 \\ -2 & 4 & 2 & 4 & -2 \\ 0 & 2 & -5 & 2 & 0 \\ -2 & 4 & 2 & 4 & -2 \\ -3 & -2 & 0 & -2 & -3 \end{bmatrix}$

6. $\begin{bmatrix} 1 & 0 & 0 & 0 & 0 \\ 0 & 1 & 0 & 0 & 0 \\ 0 & 0 & 1 & 0 & 0 \\ 0 & 0 & 0 & 1 & 0 \\ 0 & 0 & 0 & 0 & 1 \end{bmatrix}$

10. $\begin{bmatrix} 1 & -3 & 0 & -1 & 0 \\ 0 & 0 & -2 & 0 & 3 \\ 2 & 0 & 0 & 0 & 0 \\ 0 & 4 & 0 & -4 & 0 \\ 5 & 0 & -5 & 0 & 6 \end{bmatrix}$

3. $\begin{bmatrix} 3 & 6 & 5 \\ 3 & 5 & 4 \\ 1 & 1 & 1 \end{bmatrix}$

7. $\begin{bmatrix} 7 & 16 & -21 \\ 3 & 7 & -10 \\ 1 & 2 & -1 \end{bmatrix}$

11. $\begin{bmatrix} 1 & 3 & 2 \\ 0 & 0 & 5 \\ 0 & 0 & 5 \end{bmatrix}$

4. $\begin{bmatrix} 3 & 6 & -5 & 10 \\ 0 & 2 & 1 & -7 \\ 2 & 5 & -4 & 6 \\ 4 & 8 & -7 & 14 \end{bmatrix}$

8. $\begin{bmatrix} -1 & 4 & 4 & 11 \\ 2 & -5 & -2 & -5 \\ 3 & -5 & -2 & -3 \\ 1 & -2 & -1 & -2 \end{bmatrix}$

12. $\begin{bmatrix} 1 & 3 & -2 & 0 \\ 0 & 2 & 4 & 6 \\ 0 & 0 & -2 & 1 \\ 0 & 0 & 0 & 5 \end{bmatrix}$

Determine if the matrices A and B are inverses of each other. Write yes or no.

13. $A = \begin{bmatrix} 5 & 3 \\ 3 & 2 \end{bmatrix}$ and $B = \begin{bmatrix} 2 & -3 \\ -3 & 5 \end{bmatrix}$

14. $A = \begin{bmatrix} 2 & -1 & 0 \\ 1 & 2 & 4 \\ 1 & 1 & 2 \end{bmatrix}$ and $B = \frac{1}{2}\begin{bmatrix} 0 & -2 & 4 \\ -2 & -4 & 8 \\ 1 & 3 & -5 \end{bmatrix}$

15. $A = \begin{bmatrix} 6 & 4 & 0 \\ -2 & 5 & 5 \\ 0 & -1 & 2 \end{bmatrix}$ and $B = \begin{bmatrix} 1 & 0 & 6 \\ 5 & -4 & 5 \\ 2 & \frac{1}{2} & 0 \end{bmatrix}$

1.8   Solving Matrix Equations

Example 13:   Solve the matrix equation $AX = B$ for X.

$$A = \begin{bmatrix} 5 & -1 \\ 0 & 3 \end{bmatrix} \text{ and } B = \begin{bmatrix} -43 & 18 \\ 24 & 6 \end{bmatrix}$$

Step 1:   Substitute the matrices A and B into the matrix equation $AX = B$.

$$\begin{bmatrix} 5 & -1 \\ 0 & 3 \end{bmatrix} X = \begin{bmatrix} -43 & 18 \\ 24 & 6 \end{bmatrix}$$

Step 2:   Unlike solving "regular" equations, matrices cannot be divided. Instead of dividing, multiply each side of the matrix equation by the inverse of A to get X by itself. Multiply each side by A^{-1} on the left. (Multiplying must be done in the same order on both sides of the matrix equation.)

$$\begin{bmatrix} 5 & -1 \\ 0 & 3 \end{bmatrix}^{-1} \begin{bmatrix} 5 & -1 \\ 0 & 3 \end{bmatrix} X = \begin{bmatrix} 5 & -1 \\ 0 & 3 \end{bmatrix}^{-1} \begin{bmatrix} -43 & 18 \\ 24 & 6 \end{bmatrix}$$

Step 3:   Solve.

$$\begin{bmatrix} 1 & 0 \\ 0 & 1 \end{bmatrix} X = \begin{bmatrix} -7 & 4 \\ 8 & 2 \end{bmatrix} \qquad X = \begin{bmatrix} -7 & 4 \\ 8 & 2 \end{bmatrix}$$

To check, multiply AX, and the answer will be B.

Solve the matrix equations for X.

1. $\begin{bmatrix} -2 & 4 \\ 0 & 3 \end{bmatrix} X = \begin{bmatrix} 22 & -8 \\ 18 & -3 \end{bmatrix}$

4. $\begin{bmatrix} 5 & -7 \\ 10 & -1 \end{bmatrix} X = \begin{bmatrix} -11 & -38 \\ 17 & -89 \end{bmatrix}$

2. $\begin{bmatrix} 6 & -4 \\ 5 & 0 \end{bmatrix} X = \begin{bmatrix} 20 & 48 \\ 20 & 40 \end{bmatrix}$

5. $\begin{bmatrix} -1 & -5 \\ -2 & 8 \end{bmatrix} X = \begin{bmatrix} -6 & -42 \\ 24 & 78 \end{bmatrix}$

3. $\begin{bmatrix} 1 & 9 & -2 \\ 0 & 4 & 0 \\ 0 & 6 & 2 \end{bmatrix} X = \begin{bmatrix} 27 & -26 & -10 \\ 16 & -12 & 0 \\ 32 & -14 & 18 \end{bmatrix}$

6. $\begin{bmatrix} 0 & -1 & 4 \\ -8 & 5 & -1 \\ -5 & 0 & 6 \end{bmatrix} X = \begin{bmatrix} 4 & -9 & 14 \\ 10 & 45 & -69 \\ 7 & 0 & -17 \end{bmatrix}$

7. $\begin{bmatrix} -3 & 8 \\ 2 & 10 \end{bmatrix} X - 2 \begin{bmatrix} 5 & 11 \\ -1 & 7 \end{bmatrix} = \begin{bmatrix} 62 & -48 \\ 92 & -12 \end{bmatrix}$

8. $\begin{bmatrix} 6 & 1 & 0 \\ 2 & 0 & -1 \\ 0 & 0 & 4 \end{bmatrix} X + \begin{bmatrix} 1 & 5 & 4 \\ 6 & 0 & 0 \\ 4 & 7 & 0 \end{bmatrix} = \begin{bmatrix} -6 & 45 & -2 \\ 2 & 11 & -8 \\ 4 & 11 & 24 \end{bmatrix}$

1.9   Applications with Matrices

Example 14:   Find a, b, c, and d such that

$$\begin{bmatrix} 2 & -1 \\ 4 & 3 \end{bmatrix} + \begin{bmatrix} a & b \\ c & d \end{bmatrix} = \begin{bmatrix} 3 & -1 \\ 2 & 2 \end{bmatrix}$$

Step 1:   Write the equation for the four sets of corresponding elements.

$2 + a = 3$ $\qquad\qquad$ $-1 + b = -1$

$4 + c = 2$ $\qquad\qquad$ $3 + d = 2$

Step 2:   Solve each of the four equations.

$a = 1$ $\qquad\qquad$ $b = 0$

$c = -2$ $\qquad\qquad$ $d = -1$

Example 15:   Lucie and her friend Laura are shopping for a new cellular phone plan. Plan A offers 400 minutes per month for $60, plus another 200 night and weekend minutes for an extra $20. Plan B offers 500 monthly minutes for $50, plus 150 night and weekend minutes for an extra $20. The two plans can be represented in the following matrices:

$$\begin{bmatrix} \$60 & \$20 \\ 400 & 200 \end{bmatrix} = A \qquad\qquad \begin{bmatrix} \$50 & \$20 \\ 500 & 150 \end{bmatrix} = B$$

If Lucie chooses Plan A and Laura chooses Plan B, what will the total cost of their services be if they also select the night and weekend minutes, and how many total minutes will they receive? What is the average cost and minutes of the plans?

Step 1:   To determine the total of the two plans, add the two matrices together, A + B.

$$\begin{bmatrix} \$60 + \$50 & \$20 + \$20 \\ 400 + 500 & 200 + 150 \end{bmatrix} = \begin{bmatrix} \$110 & \$40 \\ 900 & 350 \end{bmatrix}$$

Step 2:   Calculate the average cost and minutes by multiplying the total matrix by $\frac{1}{2}$ (or dividing it by 2):

$$\frac{1}{2} \times \begin{bmatrix} \$110 & \$40 \\ 900 & 350 \end{bmatrix} = \begin{bmatrix} \$55 & \$20 \\ 450 & 175 \end{bmatrix}$$

Solve the following matrix problems.

1. $\begin{bmatrix} d & 3 \\ e & 1 \end{bmatrix} + \begin{bmatrix} 2 & f \\ 2 & g \end{bmatrix} = \begin{bmatrix} 5 & 3 \\ 1 & 2 \end{bmatrix}$

2. $\begin{bmatrix} 3d & 1 \\ 2 & 3g \\ 1 & f \end{bmatrix} - \begin{bmatrix} 2d & 0 \\ e & g \\ 1 & 2 \end{bmatrix} = \begin{bmatrix} -2 & 1 \\ 4 & 4 \\ 0 & 1 \end{bmatrix}$

3. $3\begin{bmatrix} 1 & 0 \\ 2 & 1 \end{bmatrix} + \dfrac{1}{2}\begin{bmatrix} 4 & 6 \\ 0 & 2 \end{bmatrix} = \begin{bmatrix} d & f \\ e & g \end{bmatrix}$

4. A computer company with one plant in the West and one plant in the East produces monitors and printers. The production for January and February are give as follows:

$$\text{January} = \begin{array}{cc} \text{West} & \text{East} \\ \text{Plant} & \text{Plant} \\ \begin{bmatrix} 2000 & 1710 \\ 800 & 650 \end{bmatrix} \end{array} \qquad \text{February} = \begin{array}{cc} \text{West} & \text{East} \\ \text{Plant} & \text{Plant} \\ \begin{bmatrix} 2300 & 1850 \\ 950 & 800 \end{bmatrix} \end{array} \begin{array}{l} \text{Monitors} \\ \text{Printers} \end{array}$$

(A) What is the average monthly production of the monitors and printers?

(B) What is the increase from January to February?

(C) What is the total production for January and February?

5. The Yummy Candy Company produces a variety of candy products and packages them for various holidays. The Christmas package consists of three pieces of chocolate, two pecan candies, one peppermint twist, and four chocolate-covered cherries. The Valentine package consists of the same package, but contains three times as many pieces of each candy. Write the number of candies in both the Christmas and Valentine packages in matrix form.

6. What are the total numbers of candies contained in one Christmas package and one Valentine package from problem 5?

1.10   Vertex-Edge Graphs

Diagrams or graphs are often helpful in visualizing networks of interconnections such as travel routes. A connectivity matrix is used to organize and analyze data representing complex interconnections.

Example 16:

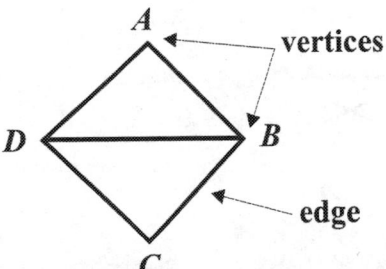

In this diagram, connecting lines indicate that there are direct routes between the two points. Conversely, if there is no line between two points, then no direct route exists between the two points. There are direct routes, also known as paths, from A to B, A to D, B to C, C to D, and B to D. There is no direct route between A and C.

This type of information can also be represented in matrix form. A zero indicates no direct route and a 1 indicates a direct route. The matrix below represents the example above.

$$
\begin{array}{c@{\ }c}
 & \begin{array}{cccc} A & B & C & D \end{array} \\
\begin{array}{c} A \\ B \\ C \\ D \end{array} &
\left[\begin{array}{cccc}
0 & 1 & 0 & 1 \\
1 & 0 & 1 & 1 \\
0 & 1 & 0 & 1 \\
1 & 1 & 1 & 0
\end{array} \right]
\end{array}
$$

The 1 in row B, column D indicates a direct route between these two locations.

Example 17:   In both the diagram and the matrix below, direct routes are indicated between A and D and between F and C. There is no direct route between A and B.

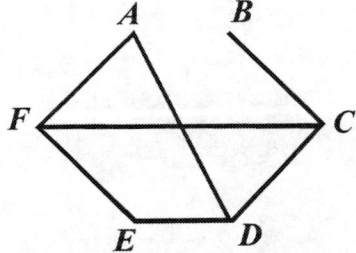

$$
\begin{array}{c@{\ }c}
 & \begin{array}{cccccc} A & B & C & D & E & F \end{array} \\
\begin{array}{c} A \\ B \\ C \\ D \\ E \\ F \end{array} &
\left[\begin{array}{cccccc}
0 & 0 & 0 & 1 & 0 & 1 \\
0 & 0 & 1 & 0 & 0 & 0 \\
0 & 1 & 0 & 1 & 0 & 1 \\
1 & 0 & 1 & 0 & 1 & 0 \\
0 & 0 & 0 & 1 & 0 & 1 \\
1 & 0 & 1 & 0 & 1 & 0
\end{array} \right]
\end{array}
$$

29

Answer each question below.

1. Write a 5×5 matrix to represent the network below. Zero indicates no direct route and one indicates a direct route.

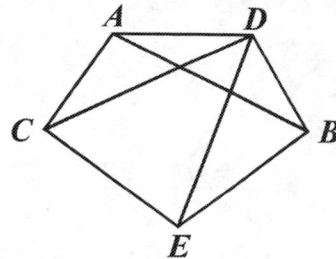

2. Which matrix represents the routes of this network?

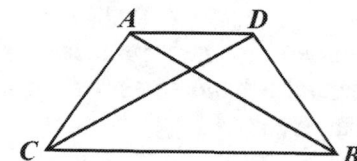

A.

$$\begin{array}{c} \\ A \\ B \\ C \\ D \end{array} \begin{array}{cccc} A & B & C & D \\ \left[\begin{array}{cccc} 0 & 1 & 1 & 1 \\ 1 & 0 & 1 & 1 \\ 1 & 1 & 0 & 1 \\ 1 & 1 & 1 & 0 \end{array}\right] \end{array}$$

B.

$$\begin{array}{c} \\ A \\ B \\ C \\ D \end{array} \begin{array}{cccc} A & B & C & D \\ \left[\begin{array}{cccc} 0 & 0 & 1 & 1 \\ 1 & 0 & 0 & 1 \\ 1 & 1 & 0 & 0 \\ 0 & 1 & 1 & 0 \end{array}\right] \end{array}$$

C.

$$\begin{array}{c} \\ A \\ B \\ C \\ D \end{array} \begin{array}{cccc} A & B & C & D \\ \left[\begin{array}{cccc} 1 & 0 & 0 & 1 \\ 0 & 0 & 1 & 1 \\ 1 & 1 & 0 & 1 \\ 0 & 1 & 1 & 0 \end{array}\right] \end{array}$$

D.

$$\begin{array}{c} \\ A \\ B \\ C \\ D \end{array} \begin{array}{cccc} A & B & C & D \\ \left[\begin{array}{cccc} 1 & 0 & 0 & 0 \\ 0 & 1 & 0 & 0 \\ 0 & 0 & 1 & 0 \\ 0 & 0 & 0 & 1 \end{array}\right] \end{array}$$

3. Draw a network represented by the matrix below.

$$\begin{array}{c} \\ A \\ B \\ C \\ D \end{array} \begin{array}{cccc} A & B & C & D \\ \left[\begin{array}{cccc} 0 & 0 & 1 & 1 \\ 0 & 0 & 1 & 1 \\ 1 & 1 & 0 & 1 \\ 1 & 1 & 1 & 0 \end{array}\right] \end{array}$$

Chapter 1 Review

Solve each matrix. If not possible, write not possible.

1. $2\begin{bmatrix} -5 & -1 \\ 4 & -7 \\ 9 & 6 \end{bmatrix} + \begin{bmatrix} -1 & -5 \\ -8 & 7 \\ 1 & -4 \end{bmatrix} =$

3. $\begin{bmatrix} 9 & -1 & -6 & 3 \\ 0 & -5 & -5 & -4 \end{bmatrix} + \begin{bmatrix} -8 & 1 & 2 & -3 \\ 1 & 6 & 11 & 5 \end{bmatrix} =$

2. $\begin{bmatrix} 4 \\ -1 \end{bmatrix} + \begin{bmatrix} -7 & -1 \end{bmatrix} =$

4. $-\dfrac{1}{4}\begin{bmatrix} -1 & -6 & 4 & -8 \\ 12 & 0 & 3 & -2 \\ 4 & 6 & -4 & 16 \end{bmatrix} =$

Subtract or multiply the following matrices. If not possible, write not possible.

5. $\begin{bmatrix} 2 & 4 \\ 0 & 5 \end{bmatrix} - \begin{bmatrix} 1 & 7 \\ -3 & 1 \end{bmatrix} =$

7. $\begin{bmatrix} -1 & 0 \\ 0 & 1 \end{bmatrix} \times \begin{bmatrix} 5 & 6 & 7 \\ 8 & 1 & 3 \end{bmatrix} =$

6. $\begin{bmatrix} 1 & 2 & 5 & 4 \\ 7 & 0 & 1 & -8 \\ -3 & 2 & 5 & -2 \end{bmatrix} - \begin{bmatrix} 1 & 3 & 13 \\ -1 & 0 & -11 \\ 0 & 5 & -7 \end{bmatrix} =$

8. $\begin{bmatrix} 1 & 2 & 5 & 4 \\ 7 & 0 & 1 & -8 \end{bmatrix} \times \begin{bmatrix} 1 & 2 & 3 & 4 \\ 2 & 3 & 4 & 1 \\ 3 & 4 & 1 & 2 \\ 4 & 3 & 2 & 1 \end{bmatrix} =$

9. Mr. Thompson goes on two road trips per year. His two favorite places to go are Dallas, TX and Atlantic City, NJ. When he went to Dallas last year, he spent $3,000 and drove 1,235 miles, and when he went to Atlantic City last year, he spent $5,500 and drove 786 miles. This year when he goes on vacation, he will have $4,300 to spend in Dallas and $4,900 in Atlantic City. He will drive 200 more miles because he is picking up his sister for both trips.

   (A) Write two 2 × 2 matrices that include miles and price for the Dallas trip and Atlantic City trip.

   (B) What is the difference in the amount of money Mr. Thompson will spend this year compared to last year?

For questions 10 and 11, find a, b, c, and d.

10. $\begin{bmatrix} a & 4 \\ b & 5 \end{bmatrix} + \begin{bmatrix} 2 & c \\ 0 & d \end{bmatrix} = \begin{bmatrix} 5 & 3 \\ 1 & 7 \end{bmatrix}$

11. $5\begin{bmatrix} 8 & 0 \\ 1 & 3 \end{bmatrix} + \dfrac{1}{2}\begin{bmatrix} 10 & 8 \\ 0 & 2 \end{bmatrix} = \begin{bmatrix} a & c \\ b & d \end{bmatrix}$

Calculate the determinant of each of the following matrices.

12. $\begin{bmatrix} -7 & 4 \\ 5 & 8 \end{bmatrix}$

14. $\begin{bmatrix} -3 & 2 \\ 20 & -4 \end{bmatrix}$

16. $\begin{bmatrix} 14 & 3 \\ -2 & 1 \end{bmatrix}$

13. $\begin{bmatrix} -9 & 9 & 5 \\ 13 & 7 & 6 \\ 2 & -6 & -7 \end{bmatrix}$

15. $\begin{bmatrix} 1 & 6 & 5 \\ -6 & -8 & -4 \\ 12 & 7 & 2 \end{bmatrix}$

17. $\begin{bmatrix} 6 & 9 & 6 \\ 7 & -8 & 5 \\ 8 & 7 & -4 \end{bmatrix}$

18. Write a 6×6 matrix to represent the network below. Zero indicates no direct route and one indicates a direct route.

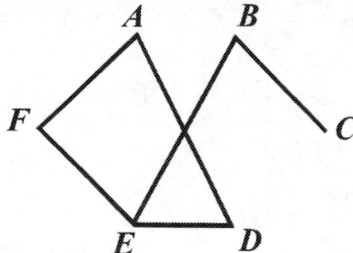

Find the inverse of each matrix below using a graphing calculator. If no inverse exists for the matrix, write none.

19. $\begin{bmatrix} 1 & 3 & -2 \\ 1 & 1 & 0 \\ 2 & 0 & 3 \end{bmatrix}$

21. $\begin{bmatrix} -1 & 3 \\ 1 & -3 \end{bmatrix}$

23. $\begin{bmatrix} 2 & 0 & 0 & 0 \\ 3 & 2 & 0 & 0 \\ -2 & 4 & -2 & 0 \\ 0 & 6 & 1 & 5 \end{bmatrix}$

20. $\begin{bmatrix} 1 & 0 & 3 & 0 \\ 0 & 2 & 0 & 4 \\ 1 & 0 & 3 & 0 \\ 0 & 2 & 0 & 4 \end{bmatrix}$

22. $\begin{bmatrix} 3 & 2 & 5 \\ 2 & 2 & 4 \\ -4 & 4 & 0 \end{bmatrix}$

24. $\begin{bmatrix} \frac{3}{5} & \frac{1}{5} \\ -\frac{2}{5} & \frac{1}{5} \end{bmatrix}$

Determine if the matrices A and B are inverses of each other. Write yes or no.

25. $A = \begin{bmatrix} 1 & 1 & 2 \\ 2 & 4 & -3 \\ 3 & 6 & -5 \end{bmatrix}$ and $B = \begin{bmatrix} 2 & -17 & 11 \\ -1 & 11 & -7 \\ 0 & 3 & -2 \end{bmatrix}$

Solve the matrix equations for X.

26. $\begin{bmatrix} 5 & 12 \\ 8 & 2 \end{bmatrix} X + \begin{bmatrix} -15 & 9 \\ 11 & 16 \end{bmatrix} = \begin{bmatrix} 97 & -300 \\ 1 & -186 \end{bmatrix}$

27. $\begin{bmatrix} 2 & 4 & 9 \\ 0 & 1 & 0 \\ 1 & 4 & 0 \end{bmatrix} X - 3 \begin{bmatrix} 1 & 4 & 0 \\ 3 & 5 & 1 \\ 3 & 9 & 0 \end{bmatrix} = \begin{bmatrix} 2 & -61 & 31 \\ -9 & -13 & 1 \\ -11 & -7 & 10 \end{bmatrix}$

Chapter 1 Test

1. $-2 \begin{bmatrix} -5 & -2 \\ 9 & -7 \\ 3 & 6 \end{bmatrix} + \begin{bmatrix} -1 & -2 \\ -8 & 7 \\ 9 & -4 \end{bmatrix} =$

A. $\begin{bmatrix} -6 & -4 \\ 1 & 0 \\ 12 & 2 \end{bmatrix}$

B. $\begin{bmatrix} -12 & -8 \\ 1 & 0 \\ 12 & 2 \end{bmatrix}$

C. $\begin{bmatrix} 9 & 2 \\ -26 & 21 \\ 3 & -16 \end{bmatrix}$

D. $\begin{bmatrix} 9 & -26 & 3 \\ 2 & 21 & -16 \end{bmatrix}$

2. $-1 \begin{bmatrix} 1 & -6 \\ 2 & -3 \end{bmatrix} + 3 \begin{bmatrix} 3 & -1 \\ 0 & -3 \end{bmatrix} =$

A. $\begin{bmatrix} 8 & 3 \\ -2 & -6 \end{bmatrix}$

B. $\begin{bmatrix} 11 & 6 \\ 1 & 9 \end{bmatrix}$

C. $\begin{bmatrix} 4 & -7 \\ 2 & -6 \end{bmatrix}$

D. $\begin{bmatrix} 10 & -9 \\ 2 & -12 \end{bmatrix}$

3. If a 2×4 matrix was multiplied by a 4×2 matrix, the resulting matrix would be what type of matrix?

A. 2×8
B. 2×4
C. 8×8
D. 2×2

4. The local hardware store has four different style grills for the summer sale. The price of the four grills are listed from least to greatest, and they are represented in the 2×2 matrix below.

$\begin{bmatrix} \$120 & \$155 \\ \$160 & \$230 \end{bmatrix}$

If the store is having a sale where everything is 20% off, what is the new price of the grill that is the second most expensive?

A. $32
B. $124
C. $184
D. $128

5. The area of a triangle with vertices at the points (a, b), (c, d), and (e, f) is $\frac{1}{2}|\det A|$, where $A = \begin{bmatrix} a & b & 1 \\ c & d & 1 \\ e & f & 1 \end{bmatrix}$.

If $\det A = ad + be + cf - ed - af - bc$ and the vertices of a triangle are at the points $(-7, 5)$, $(3, -6)$, and $(1, 9)$, what is the area of the triangle?

A. -64
B. 64
C. 32
D. 128

6. What is the inverse of $\begin{bmatrix} -4 & 7 \\ 6 & -10 \end{bmatrix}$?

A. $\begin{bmatrix} -10 & -7 \\ -6 & -4 \end{bmatrix}$

B. $\begin{bmatrix} 10 & 7 \\ 6 & 4 \end{bmatrix}$

C. $\begin{bmatrix} -5 & -\frac{7}{2} \\ -3 & -2 \end{bmatrix}$

D. $\begin{bmatrix} 5 & \frac{7}{2} \\ 3 & 2 \end{bmatrix}$

7. The discount store next to the mall is having a huge sale. All pants normally priced $25.95 are now 40% off, and all shoes normally priced $14.95 are now 60% off. Which of the following best represents this information in a matrix?

A. $\begin{bmatrix} \$25.95 & 40\% \\ \$14.95 & 60\% \end{bmatrix}$

B. $\begin{bmatrix} \$25.95 & 60\% \\ \$14.95 & 40\% \end{bmatrix}$

C. $\begin{bmatrix} \$25.95 & \$14.95 \\ 40\% & 60\% \end{bmatrix}$

D. Both A and C represent the data correctly.

8. $\begin{bmatrix} 1 & 3 \\ 1 & -4 \end{bmatrix} + 3\begin{bmatrix} 3 & -1 & 0 \\ 6 & 5 & 2 \end{bmatrix} =$

A. $\begin{bmatrix} 10 & 0 & 0 \\ 19 & 11 & 6 \end{bmatrix}$

B. $\begin{bmatrix} 10 & -3 & 0 \\ 19 & 15 & 2 \end{bmatrix}$

C. $\begin{bmatrix} 10 & 0 \\ 19 & 11 \end{bmatrix}$

D. Not possible

9. $\begin{bmatrix} 0 & -2 \\ 4 & 5 \end{bmatrix} + \begin{bmatrix} -2 & -1 \\ 6 & 14 \end{bmatrix} =$

A. $\begin{bmatrix} -2 & 1 \\ 2 & 9 \end{bmatrix}$

B. $\begin{bmatrix} -2 & 2 \\ 24 & 5 \end{bmatrix}$

C. $\begin{bmatrix} -2 & -3 \\ 10 & 19 \end{bmatrix}$

D. $\begin{bmatrix} 2 & 3 \\ 10 & 19 \end{bmatrix}$

10. $\frac{1}{2}\begin{bmatrix} 5 & 8 & -4 \\ -1 & 12 & 6 \end{bmatrix} =$

A. $\begin{bmatrix} \frac{5}{2} & 8 & -4 \\ -1 & 12 & 6 \end{bmatrix}$

B. $\begin{bmatrix} 10 & 16 & -8 \\ -2 & 24 & 12 \end{bmatrix}$

C. $\begin{bmatrix} \frac{5}{2} & 4 & -2 \\ -\frac{1}{2} & 6 & 3 \end{bmatrix}$

D. $\begin{bmatrix} \frac{5}{2} & 4 & -2 \\ -1 & 12 & 6 \end{bmatrix}$

11. $\begin{bmatrix} 2 & 4 & 6 \\ 8 & 0 & -2 \end{bmatrix} \times \begin{bmatrix} -6 \\ 7 \\ 8 \end{bmatrix} =$

A. $\begin{bmatrix} -64 \\ 64 \end{bmatrix}$

B. $\begin{bmatrix} 64 \\ -64 \end{bmatrix}$

C. $\begin{bmatrix} 98 \\ -64 \end{bmatrix}$

D. $\begin{bmatrix} 98 \\ 64 \end{bmatrix}$

12. $\begin{bmatrix} 1 & 3 \\ 0 & -1 \end{bmatrix} \times \begin{bmatrix} 4 & -1 \\ 2 & 0 \end{bmatrix} =$

A. $\begin{bmatrix} 10 & -1 \\ -2 & 0 \end{bmatrix}$

B. $\begin{bmatrix} -2 & -1 \\ 2 & 0 \end{bmatrix}$

C. $\begin{bmatrix} 1 & 1 \\ 0 & 0 \end{bmatrix}$

D. $\begin{bmatrix} 10 & -1 \\ 0 & 0 \end{bmatrix}$

13. What is the inverse of $\begin{bmatrix} -3 & 8 \\ 2 & -5 \end{bmatrix}$?

    A. $\begin{bmatrix} 5 & 8 \\ 2 & 3 \end{bmatrix}$

    B. $\begin{bmatrix} 3 & -8 \\ -2 & 5 \end{bmatrix}$

    C. $\begin{bmatrix} -5 & 8 \\ 2 & -3 \end{bmatrix}$

    D. $\begin{bmatrix} 1 & 0 \\ 0 & 1 \end{bmatrix}$

14. What is the inverse of $\begin{bmatrix} 2 & 5 & 4 \\ 1 & 3 & 3 \\ -2 & 0 & 8 \end{bmatrix}$?

    A. $\begin{bmatrix} -2 & -5 & -4 \\ -1 & -3 & -3 \\ 2 & 0 & -8 \end{bmatrix}$

    B. $\begin{bmatrix} 4 & 3 & 8 \\ 5 & 3 & 0 \\ 2 & 1 & -2 \end{bmatrix}$

    C. $\begin{bmatrix} 12 & -20 & 1.5 \\ -7 & 12 & -1 \\ 3 & -5 & 0.5 \end{bmatrix}$

    D. $\begin{bmatrix} 0.5 & -1 & 1.5 \\ -5 & 12 & -20 \\ 3 & -7 & 12 \end{bmatrix}$

15. What is the inverse of $\begin{bmatrix} 4 & -10 \\ -2 & 5 \end{bmatrix}$?

    A. $\begin{bmatrix} -4 & 10 \\ 2 & -5 \end{bmatrix}$

    B. $\begin{bmatrix} 5 & 10 \\ 2 & 4 \end{bmatrix}$

    C. $\begin{bmatrix} 2.5 & 5 \\ 1 & 2 \end{bmatrix}$

    D. No inverse exists for this matrix.

16. Solve the equation $AX = B$ for X.
$$A = \begin{bmatrix} -1 & 5 \\ 0 & 6 \end{bmatrix} \text{ and } B = \begin{bmatrix} 43 & 1 \\ 54 & 0 \end{bmatrix}$$

    A. $X = \begin{bmatrix} 2 & -1 \\ 9 & 0 \end{bmatrix}$

    B. $X = \begin{bmatrix} 227 & -1 \\ 324 & 0 \end{bmatrix}$

    C. $X = \begin{bmatrix} -43 & 36 \\ -54 & 45 \end{bmatrix}$

    D. $X = \begin{bmatrix} -43 & \frac{1}{5} \\ 0 & 0 \end{bmatrix}$

Chapter 2
Systems of Equations and Systems of Inequalities

This chapter covers the following Georgia Performance Standards:

| MM3A | Algebra | MM3A5a, MM3A5c |
| | | MM3A6a, MM3A6b |

We call two linear equations considered at the same time a **system** of linear equations. The graph of a linear equation is a straight line. The graphs of two linear equations can show that the lines are **parallel**, **intersecting**, or **collinear**. Two lines that are **parallel** will never intersect and have no ordered pairs in common. If two lines are **intersecting**, they have one point in common, and in this chapter, you will learn to find the ordered pair for that point. If the graph of two linear equations is the same line, we say the lines are **collinear**.

If you are given a system of two linear equations, and you put both equations in slope-intercept form, you can immediately tell if the graph of the lines will be **parallel**, **intersecting**, or **collinear**.

If two linear equations have the same slope and the same y-intercept, then they are both equations for the same line. They are called **collinear** or **coinciding** lines. A line is made up of an infinite number of points extending infinitely far in two directions. Therefore, collinear lines have an infinite number of points in common.

Example 1: $2x + 3y = -3$ **In slope intercept form:** $y = -\dfrac{2}{3}x - 1$

$4x + 6y = -6$ **In slope intercept form:** $y = -\dfrac{2}{3}x - 1$

The slope and y-intercept of both lines are the same.

If two linear equations have the same slope but different y-intercepts, they are **parallel** lines. Parallel lines never touch each other, so they have no points in common.

If two linear equations have different slopes, then they are intersecting lines and share exactly one point in common.

The chart below summarizes what we know about the graphs of two equations in slope-intercept form.

y-Intercepts	Slopes	Graphs	Number of Solutions
same	same	collinear	infinite
different	same	distinct parallel lines	none (they never touch)
same or different	different	intersecting lines	exactly one

For the pairs of equations below, put each equation in slope-intercept form, and tell whether the graphs of the lines will be collinear, parallel, or intersecting.

1. $x - y = -1$
    $-x + y = -1$

2. $x - 2y = 4$
    $-x + 2y = 6$

3. $y - 2 = x$
    $x + 2 = y$

4. $x = y - 1$
    $-x = y - 1$

5. $2x + 5y = 10$
    $4x + 10y = 20$

6. $x + y = 3$
    $x - y = 1$

7. $2y = 4x - 6$
    $-6x + y = 3$

8. $x + y = 5$
    $2x + 2y = 10$

9. $2x = 3y - 6$
    $4x = 6y - 6$

10. $2x - 2 = 2$
    $3y = -x + 5$

11. $x = -y$
    $x = 4 - y$

12. $2x = y$
    $x + y = 3$

13. $x = y + 1$
    $y = x + 1$

14. $x - 2y = 4$
    $-2x + 4y = -8$

15. $2x + 3y = 4$
    $-2x + 3y = -8$

16. $2x - 4y = 1$
    $-6x + 12y = 3$

17. $-3x + 4y = 1$
    $6x + 8y = 2$

18. $x + y = 2$
    $5x + 5y = 10$

19. $x + y = 4$
    $x - y = 4$

20. $y = -x + 3$
    $x - y = 1$

2.1   Finding Common Solutions for Intersecting Lines

When two lines intersect, they share exactly one point in common.

Example 2: $3x + 4y = 20$ and $2y - 4x = 12$

Put each equation in slope-intercept form.

$$
\begin{array}{ll}
3x + 4y = 20 & \qquad 2y - 4x = 12 \\
4y = -3x + 20 & \qquad 2y = 4x + 12 \\
y = -\frac{3}{4}x + 5 & \qquad y = 2x + 6
\end{array}
$$

slope-intercept form

Straight lines with different slopes are **intersecting lines**. Look at the graphs of the lines on the same Cartesian plane.

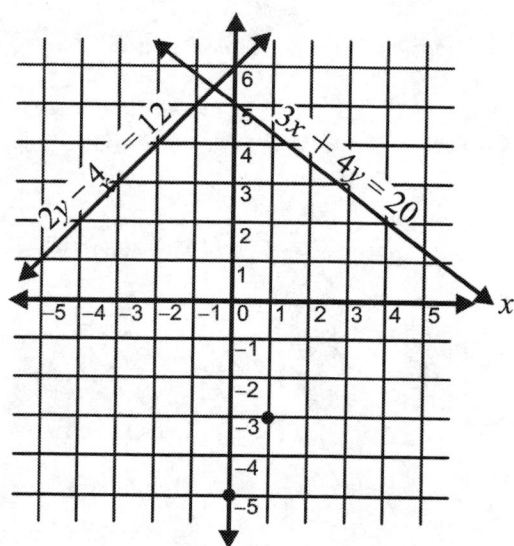

You can see from looking at the graph that the intersecting lines share one point in common. However, it is hard to tell from looking at the graph what the coordinates are for the point of intersection. To find the exact point of intersection, you can use the **substitution method** to solve the system of equations algebraically.

2.2   Solving Systems of Equations by Substitution

You can solve systems of equations by using the substitution method.

Example 3:   Find the point of intersection of the following two equations:

Equation 1: $x - y = 3$

Equation 2: $2x + y = 9$

Step 1:   Solve one of the equations for x or y. Let's choose to solve equation 1 for x.

Equation 1: $x - y = 3$

$x = y + 3$

Step 2:   Substitute the value of x from equation 1 in place of x in equation 2.

Equation 2: $2x + y = 9$
$2(y + 3) + y = 9$
$2y + 6 + y = 9$
$3y + 6 = 9$
$3y = 3$
$y = 1$

Step 3:   Substitute the solution for y back in equation 1 and solve for x.

Equation 1: $x - y = 3$
$x - 1 = 3$
$x = 4$

Step 4:   The solution set is $(4, 1)$. Substitute $(4, 1)$ in both of the equations to check.

Equation 1: $x - y = 3$ Equation 2: $2x + 9 = 9$
$4 - 1 = 3$ $2(4) + 1 = 9$
$3 = 3$ $8 + 1 = 9$
$9 = 9$

The point $(4, 1)$ is common for both equations. This is the **point of intersection**.

For each of the following pairs of equations, find the point of intersection using the substitution method.

1. $x + 2y = 8$
    $2x - 3y = 2$

2. $x - y = -5$
    $x + y = 1$

3. $x - y = 4$
    $x + y = 2$

4. $x - y = -1$
    $x + y = 9$

5. $-x + y = 2$
    $x + y = 8$

6. $x + 4y = 10$
    $x + 5y = 10$

7. $2x + 3y = 2$
    $4x - 9y = -1$

8. $x + 3y = 5$
    $x - y = 1$

9. $-x = y - 1$
    $x = y - 1$

10. $x - 2y = 2$
    $2y + x = -2$

11. $5x + 2y = 1$
    $2x + 4y = 10$

12. $3x - y = 2$
    $5x + y = 6$

13. $2x + 3y = 3$
    $4x + 5y = 5$

14. $x - y = 1$
    $-x - y = 1$

15. $x = y + 3$
    $y = 3 - x$

2.3   Solving Systems of Equations by Adding or Subtracting

You can solve systems of equations algebraically by adding or subtracting an equation from another equation or system of equations.

Example 4:   Find the point of intersection of the following two equations:
Equation 1: $x + y = 10$
Equation 2: $-x + 4y = 5$

Step 1:   Eliminate one of the variables by adding the two equations together. Since the x has the same coefficient in each equation, but opposite signs, it will cancel nicely by adding.

$$x + y = 10$$
$$\underline{+(-x + 4y = 5)} \quad \text{Add each like term together.}$$
$$0 + 5y = 15 \quad \text{Simplify.}$$
$$5y = 15 \quad \text{Divide both sides by 5.}$$
$$y = 3$$

Step 2:   Substitute the solution for y back into an equation, and solve for x.

Equation 1: $x + y = 10$ Substitute 3 for y.
$x + 3 = 10$ Subtract 3 from both sides.
$x = 7$

Step 3:   The solution set is $(7, 3)$. To check, substitute the solution into both of the original equations.

Equation 1: $x + y = 10$ Equation 2: $-x + 4y = 5$
$7 + 3 = 10$ $-(7) + 4(3) = 5$
$10 = 10$ $-7 + 12 = 5$
$5 = 5$

The point $(7, 3)$ is the point of intersection.

Example 5:   Find the point of intersection of the following two equations:
Equation 1: $3x - 2y = -1$
Equation 2: $-4y = -x - 7$

Step 1:   Put the variables in equation 2 on the same side.
$-4y = -x - 7$ Add x to both sides.
$x - 4y = -x + x - 7$ Simplify.
$x - 4y = -7$

Step 2:   Add the two equations together to cancel one variable. Since each variable has the same sign and different coefficients, we have to multiply one equation by a negative number so one of the variables will cancel. Equation 1's y variable has a coefficient of 2, and if multiplied by -2, the y will have the same variable as the y in equation 2, but a different sign. This will cancel nicely when added.
$-2(3x - 2y = -1)$ Multiply by -2.
$-6x + 4y = 2$

Step 3:   Add the two equations.
$$\begin{array}{r} -6x + 4y = 2 \\ + (x - 4y = -7) \\ \hline -5x + 0 = -5 \end{array}$$ Add equation 2 to equation 1.
Simplify.
$-5x = -5$ Divide both sides by -5.
$x = 1$

Step 4:   Substitute the solution for x back into an equation and solve for y.
Equation 1: $3x - 2y = -1$ Substitute 1 for x.
$3(1) - 2y = -1$ Simplify.
$3 - 2y = -1$ Subtract 3 from both sides.
$3 - 3 - 2y = -1 - 3$ Simplify.
$-2y = -4$ Divide both sides by -2.
$y = 2$

Step 5:   The solution set is $(1, 2)$. To check, substitute the solution into both of the original equations.

Equation 1: $3x - 2y = -1$ Equation 2: $-4y = -x - 7$
$3(1) - 2(2) = -1$ $-4(2) = -1 - 7$
$3 - 4 = -1$ $-8 = -8$
$-1 = -1$

The point $(1, 2)$ is the point of intersection.

For each of the following pairs of equations, find the point of intersection by adding the two equations together.

1. $x + 2y = 8$
   $-x - 3y = 2$

2. $x - y = 5$
   $2x + y = 1$

3. $x - y = -1$
   $x + y = 9$

4. $3x - y = -1$
   $x + y = 13$

5. $-x + 4y = 2$
   $x + y = 8$

6. $x + 4y = 10$
   $x + 7y = 16$

7. $2x - y = 2$
   $4x - 9y = -3$

8. $x + 3y = 13$
   $5x - y = 1$

9. $-2x = 2y - 2$
   $5x = 5y - 5$

10. $x - y = 2$
    $2y + x = 5$

11. $5x + 2y = 1$
    $4x + 8y = 20$

12. $3x - 2y = 14$
    $x - y = 6$

13. $2x + 3y = 3$
    $3x + 5y = 5$

14. $x - 4y = 6$
    $-x - y = -1$

15. $x = 2y + 3$
    $y = 3 - x$

2.4   Solving Systems of Equations Using Matrices

There are many different ways to solve systems of equations. Using matrices is one way to solve equations. For this section, a graphing calculator will be needed to help determine solutions.

Example 6:   Given the equations $x + y = 5$ and $3x - y = -1$, create two matrices and use a calculator to find the solution.

Step 1:   Write the equations so that the corresponding variables line up.

Equation 1: $x \; + \; y \; = \; 5$
Equation 2: $3x \; - \; y \; = \; -1$

Step 2:   Place the coefficients of x and y into matrix A (the coefficient matrix).
The coefficients of x will go in the first column and the coefficients of y will go into the second column.
The top row contains the coefficients of x and y of equation 1, and the bottom row contains coefficients of x and y of equation 2.

$$A = \begin{bmatrix} 1 & 1 \\ 3 & -1 \end{bmatrix}$$

To put this matrix into your calculator, perform the following steps. (This is based on a TI-83 calculator, but other calculators are similar.)
1. Find and press the MATRIX key.
2. Go over to the right until you have the word EDIT highlighted.
3. We want to edit matrix A, so press ENTER.
4. This is a 2×2 matrix, so type 2, enter, 2, enter.
5. Put the coefficients into the matrix in the same order as above. (You will type 1, enter, 1, enter, 3, enter, -1, enter.)
We are done with this matrix.

Step 3:   Place the constants into matrix B.
The top row is the constant for equation 1, and the bottom row is the constant for equation 2.

$$B = \begin{bmatrix} 5 \\ -1 \end{bmatrix}$$

To put this matrix into your calculator, perform the following steps. (This is based on a TI-83 calculator, but other calculators are similar.)
1. Find and press the MATRIX key.
2. Go over to the right until you have the word EDIT highlighted.
3. We want to edit matrix B, so go down to B and press ENTER.
4. This is a 2×1 matrix, so type 2, enter, 1, enter.
5. Put the constants into the matrix in the same order as above. (You will type 5, enter, -1, enter.)
We are done with this matrix. To exit, press 2nd QUIT to get back to the main screen.

Step 4:   The next step is to find the solution of the system by multiplying the inverse of matrix A to matrix B. This step will be performed using a graphing calculator.
1. Find and press the MATRIX key.
2. Press enter. This should show [A] on your screen.
3. Find the x^{-1} key. (This performs the inverse function.) Press the x^{-1} key.
4. Your screen should now look like $[A]^{-1}$.
5. Find and press the MATRIX key again.
6. Go down to [B] and press enter.
7. Your screen should now look like $[A]^{-1}[B]$.
8. Press enter. Your calculator should give you the 2×1 matrix $\begin{bmatrix} 1 \\ 4 \end{bmatrix}$ as the answer.

The 1 is your x-value and the 4 is your y-value. $(1, 4)$ is the solution to the system of equations.

Use matrices and a calculator to solve the system of equations.

1. $-3x + y = -2$
   $x + y = -6$

2. $x - 2y = 14$
   $x + 3y = 9$

3. $5x + y = 13$
   $3x + 3y = 15$

4. $4x - y = 10$
   $2x + 3y = 12$

5. $2x + 4y = 36$
   $10y - 5x = 0$

6. $2x + 4y = 3$
   $x + 3y = 13$

7. $-3x + 2y = 7$
   $5x - 4y = -12$

8. $4x + 3y = -1$
   $5x + 4y = 1$

9. $3x + 4y = 25$
   $4x - 7y = -16$

10. $2x + 2y = 8$
    $-2x - 3y = 2$

11. $x - 5y = 5$
    $2x + 10y = 6$

12. $x - y = -1$
    $2x + 2y = 18$

13. $3x - y = -1$
    $3x + 3y = 39$

14. $-\frac{1}{2}x + 2y = 1$
    $x + y = 8$

15. $x + 4y = 10$
    $-x - 7y = -16$

16. $4x - 2y = 4$
    $4x - 9y = -3$

17. $x + 3y = 13$
    $10x - 2y = 2$

18. $-2x = 2y - 2$
    $x = y - 1$

2.5   Graphing Systems of Inequalities

We solve systems of inequalities best graphically. Look at the following example.

Example 7:   Sketch the solution set of the following system of inequalities:

$$y > -2x - 1 \text{ and } y \leq 3x$$

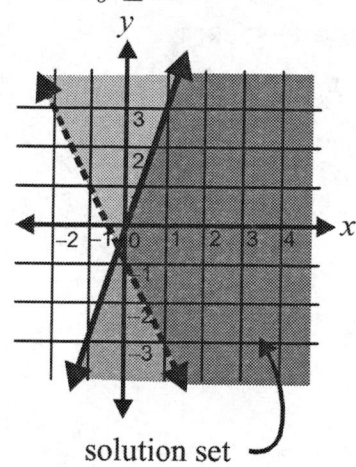

solution set

Step 1:   Graph both inequalities on a Cartesian plane.

Step 2:   Shade the portion of the graph that represents the solution set to each inequality.

Step 3:   Any shaded region that overlaps is the solution set of both inequalities.

Graph the following systems of inequalities on your own graph paper. Shade and identify the solution set for both inequalities.

1. $2x + 2y \geq -4$
    $3y < 2x + 6$

2. $7x + 7y \leq 21$
    $8x < 6y - 24$

3. $9x + 12y < 36$
    $34x - 17y > 34$

4. $-11x - 22y \geq 44$
    $-4x + 2y \leq 8$

5. $24x < 72 + 36y$
    $11x + 22y \leq -33$

6. $15x - 60 < 30y$
    $20x + 10y < 40$

7. $-12x + 24y > -24$
    $10x < -5y + 15$

8. $y \geq 2x + 2$
    $y < -x - 3$

9. $3x + 4y \geq 12$
    $y > -3x + 2$

10. $-3x \leq 6 + 2y$
    $y \geq -x - 2$

11. $2x - 2y \leq 4$
    $3x + 3y \leq -9$

12. $-x \geq -2y - 2$
    $-2x - 2y > 4$

2.6   Solving Word Problems with Systems of Equations

Certain word problems can be solved using systems of equations.

Example 8:   In a game show, Andre earns 6 points for every right answer and loses 12 points for every wrong answer. He has answered correctly 12 times as many as he has missed. His final score was 120. How many times did he answer correctly?

Step 1:   Let r = number of right answers.
Let w = number of wrong answers.

We know 2 sets of information that can be made into equations with 2 variables.

He earns +6 points for right answers and loses 12 points for wrong answers.

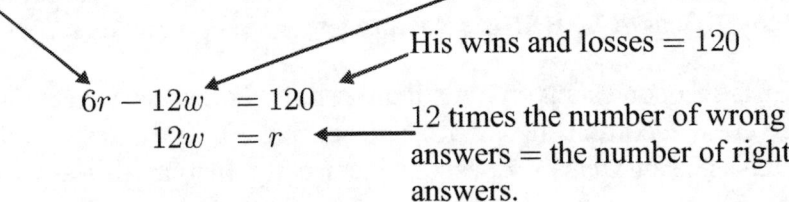

His wins and losses = 120

$$6r - 12w = 120$$
$$12w = r$$

12 times the number of wrong answers = the number of right answers.

Step 2:   Substitute the value for r $(12w)$ in the first equation.

$$6(12w) - 12w = 120$$
$$72w - 12w = 120$$
$$60w = 120$$
$$w = 2$$

Step 3:   Substitute the value for w back in the equation.

$$6r - 12(2) = 120$$
$$6r - 24 = 120$$
$$6r = 144$$
$$r = 24$$

Use systems of equations to solve the following word problems.

1. The sum of two numbers is 140 and their difference is 20. What are the two numbers?

2. The sum of two numbers is 126 and their difference is 42. What are the two numbers?

3. Kayla gets paid $6.00 for raking leaves and $8.00 for mowing each lawn of the neighbors around her subdivision. This year she mowed the lawns 12 times more than she raked leaves. In total, she made $918.00 for doing both. How many times did she rake the leaves?

4. Prices for the movie are $4.00 for children and $8.00 for adults. The total amount of ticket sales is $1176. There are 172 tickets sold. How many adults and children bought tickets?

5. A farmer sells a dozen eggs at the market for $2.00 and one of his bags of grain for $5.00. He has sold 5 times as many bags of grain as he has dozens of eggs. By the end of the day, he has made $243.00 worth of sales. How many bags of grain did he sell?

6. Every time Lauren does one of her chores, she gets 15 minutes to talk on the phone. When she does not perform one of her chores, she gets 20 minutes of phone time taken away. This week she has done her chores 5 times more than she has not performed her chores. In total, she has accumulated 165 minutes. How many times has Lauren not performed her chores?

7. The choir sold boxes of candy and teddy bears near Valentine's Day to raise money. They sold twice as many boxes of candy as they did teddy bears. Bears sold for $8.00 each and candy sold for $6.00. They collected $380. How much of each item did they sell?

8. Mr. Marlow keeps ten and twenty dollar bills in his dresser drawer. He has 1 less than twice as many 10's as 20's. He has $550 altogether. How many tens does he have?

9. Kosta is a contestant on a math quiz show. For every correct answer, Kosta receives $18.00. For every incorrect answer, Kosta loses $24.00. Kosta answers the questions correctly twice as often as he answers the questions incorrectly. In total, Kosta wins $72.00. How many questions does Kosta answer incorrectly?

10. John Vasilovik works in landscaping. He gets paid $50 for each house he pressure-washes and $20 for each lawn he mows. He gets 4 times more jobs for mowing lawns than for pressure-washing houses. During a given month, John earns $2600. How many houses does John pressure wash?

11. Every time Stephen walks the dog, he gets 30 minutes to play video or computer games. When he does not take out the dog on time, he gets a mess to clean up and loses 1 hour of video/computer game time. This week he has walked the dog on time 8 times more than he did not walk the dog on time. In total, he has accumulated 3 hours of video/computer time. How many times has Stephen not walked the dog on time?

12. On Friday, Rosa bought party hats and kazoos for her friend's birthday party. On Saturday she decided to purchase more when she found out more people were coming. How much did she pay for each party hat?

	Hats	Kazoos	Total Cost
Friday	15	20	$15.00
Saturday	10	5	$8.75

13. Timothy and Jesse went to purchase sports clothing they needed as soccer players. The table below shows what they bought and the amount they paid. What is the price of 1 soccer jersey?

	Soccer Jerseys	Tube Socks	Total Cost
Timothy	4	7	$78.30
Jesse	3	5	$57.60

2.7   Solving Problems Using Linear Programming

Just like you learned in the previous section, you can solve realistic problems involving systems of equations using your graphing calculator and matrices.

Example 9:   Marquise and Christene go to Depot Max to go back-to-school supply shopping. Marquise buys 7 binders and 3 packs of pencils for $34.92. Christene buys the same brand of binders and pencils, but buys only 4 binders and 4 packs of pencils for $23.68. How much does a single binder and a single pack of pencils cost?

   Step 1:   First, set up the system of equations from the information given in the problem. Assign binders to x and packs of pencils to y.

$$7x + 3y = 34.92$$
$$4x + 4y = 23.68$$

   Step 2:   Now from the knowledge about putting equations into matrices in calculators, plug in the information.

$$A = \begin{bmatrix} 7 & 3 \\ 4 & 4 \end{bmatrix} \qquad B = \begin{bmatrix} 34.92 \\ 23.68 \end{bmatrix}$$

   Step 3:   Remember, when solving the equations you must multiply the inverse of matrix A to matrix B.

$$[A]^{-1}[B]$$

   Answer: The calculator should give you an answer in a 2×1 matrix with it being $\begin{bmatrix} 4.29 \\ 1.63 \end{bmatrix}$ and each binder costs $4.29 and each pack of pencils costs $1.63.

Use your graphing calculator to put the following realistic problems into matrices and solve.

1. Two customers visit the same gas station. One customer buys 15 gallons of gasoline and buys 2 bottles of soda for a total purchase of $42.73. The other customer buys the same type of gas, but only 8 gallons of gasoline and buys the same type of soda, but only 1 bottle for a total purchase of $22.71. How much is one gallon of gasoline and one bottle of soda?

2. Steve buys 6 books and 3 sets of flash cards for a total of $224.67 and Stephanie buys 10 books and 4 sets of flash cards for a total of $359.46. How much is each book and set of flash cards?

3. Chris buys 2 pounds of apples and 2 pounds of oranges for a total of $9.76. Becky buys the same type of apples and oranges, but instead buys 4 pounds of apples and 1 pound of oranges for a total of $10.64. How much is one pound of each the apples and oranges?

4. The sum of two consecutive odd integers is 240. If the lesser odd integer is doubled, then the new sum is 359. What are the two integers?

Chapter 2 Review

For each pair of equations below, tell whether the graphs of the lines will be collinear, parallel, or intersecting.

1. $y = 4x + 1$
   $y = 4x - 3$

2. $x + y = 5$
   $x - y = -1$

3. $5y = 3x - 7$
   $4x - 3y = -7$

Find the common solution for each of the following pairs of equations, using the substitution method.

4. $x - y = 2$
   $x + 4y = -3$

6. $-4y = -2x + 4$
   $x = -2y - 2$

8. $x = y - 3$
   $-x = y + 3$

5. $x + y = 1$
   $x + 3y = 1$

7. $2x + 8y = 20$
   $5y = 12 - x$

9. $-2x + y = -3$
   $x - y = 9$

Graph the following systems of inequalities on your own graph paper. Shade the solution set to both inequalities.

10. $x + 2y \geq 2$
    $2x - y \leq 4$

12. $6x + 8y \leq -24$
    $-4x + 8y \geq 16$

14. $2y \geq 6x + 6$
    $2x - 4y \geq -4$

11. $20x + 10y \leq 40$
    $3x + 2y \geq 6$

13. $14x - 7y \geq -28$
    $3x + 4y \leq -12$

15. $9x - 6y \geq 18$
    $3y \geq 6x - 12$

Find the point of intersection for each pair of equations by adding and/or subtracting the two equations.

16. $2x + y = 4$
    $3x - y = 6$

18. $x + y = 1$
    $y = x + 7$

20. $2x - 2y = 7$
    $3x - 5y = \frac{5}{2}$

17. $x + 2y = 3$
    $x + 5y = 0$

19. $2x + 4y = 5$
    $3x + 8y = 9$

21. $x - 3y = -2$
    $y = -\frac{1}{3}x + 4$

Use matrices and a calculator to solve each system of equations.

22. $3x + y = 8$
    $2x + y = 4$

24. $x = y + 1$
    $y = 6x + 4$

26. $y = 7x + 4$
    $3x - 2y = 11$

23. $\frac{1}{4}x + 6y = 1$
    $x - 2y = 4$

25. $y = \frac{1}{2}x + 10$
    $5x - 6y = 16$

27. $3x + y = 7$
    $7x - 6y = 0$

Use systems of equations to solve the following word problems.

28. Hargrove High School sold 227 tickets for their last basketball game. Adult tickets sold for $5 and student tickets were $2. How many adult tickets were sold if the ticket sales totaled $574?

29. Zack is an ostrich and llama breeder. He sells full-grown ostriches for $625 and full-grown llamas for $750 each. Zack sold 1 less than 3 times as many llamas as ostriches this year. His total sales for the year were $7, 875.00. How many llamas did Zack sell during this year?

30. Sarah and Abdul played Geography Quiz Bowl during summer school. For every time Abdul got an answer right, Sarah got 4 answers right. If Sarah and Abdul correctly answered 75 questions, how many times did Abdul answer correctly?

Chapter 2 Test

1. The graphs of the equations $x = -y$ and $x = 4 - y$ are

   A. collinear.
   B. parallel.
   C. intersecting.
   D. not enough information.

2. The graphs of the equations $x + y = 2$ and $5x + 5y = 10$ are

   A. collinear.
   B. parallel.
   C. intersecting.
   D. not enough information.

3. What is the intersection point of the graphs of the equations $x = y + 3$ and $y = 3 - x$?

   A. $(3, 0)$
   B. $(0, 3)$
   C. $(3, 3)$
   D. $(-3, 0)$

4. What is the intersection point of the graphs of the equations $2x + 3y = 2$ and $4x - 9y = -1$?

   A. $(2, 3)$

   B. $\left(\dfrac{1}{3}, \dfrac{1}{2}\right)$

   C. $\left(\dfrac{1}{2}, \dfrac{1}{3}\right)$

   D. $(3, 2)$

5. What is the solution of the graphs of the equations $-x = y - 1$ and $x = y - 1$?

   A. $(0, 1)$
   B. $(1, 0)$
   C. $(-2, -1)$
   D. $(2, 1)$

6. What is the solution of the graphs of the equations $2x - y = 2$ and $4x - 9y = -3$?

   A. $\left(\dfrac{1}{2}, -1\right)$

   B. $\left(1, \dfrac{3}{2}\right)$

   C. $\left(-\dfrac{3}{2}, 1\right)$

   D. $\left(\dfrac{3}{2}, 1\right)$

7. At what point do the graphs of the equations $x - 4y = 6$ and $-x - y = -1$ intersect?

   A. $(2, 1)$
   B. $(2, -1)$
   C. $(-2, 1)$
   D. $(-2, -1)$

8. The admission fee at the fair is $1.50 for children and $4 for adults. On a certain day, $2,200$ people enter the fair and $5,050$ is collected. How many children attended? How many adults attended?

   A. children $= 750$, adults $= 1000$
   B. children $= 1000$, adults $= 700$
   C. children $= 1300$, adults $= 750$
   D. children $= 1500$, adults $= 700$

9. Which graph represents $2x + 2y \geq -4$ and $3y < 2x + 6$?

10. Which graph represents $2x - 2y \leq 4$ and $3x + 3y \leq -9$?

A.

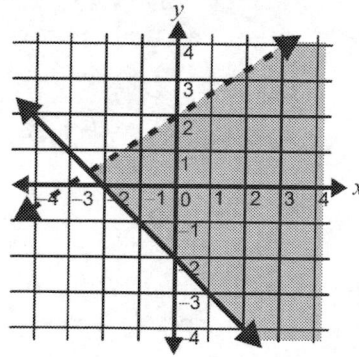

A.

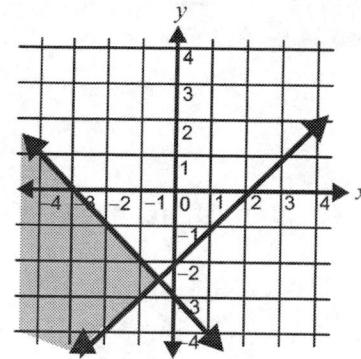

B.

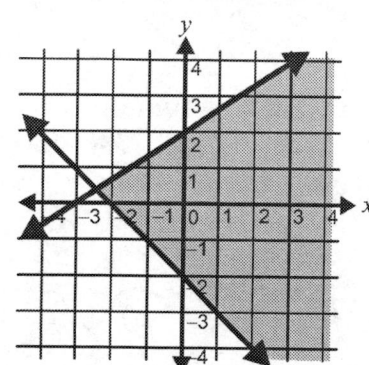

B.

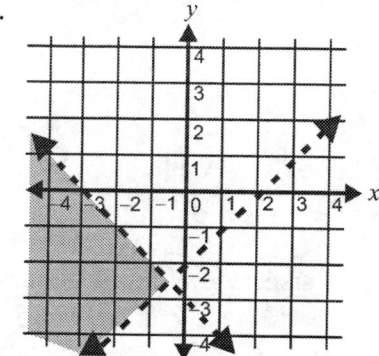

C.

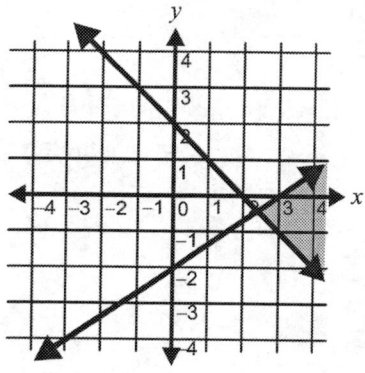

C.

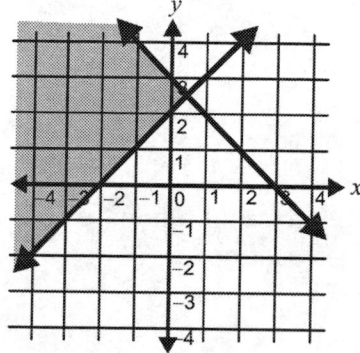

D.

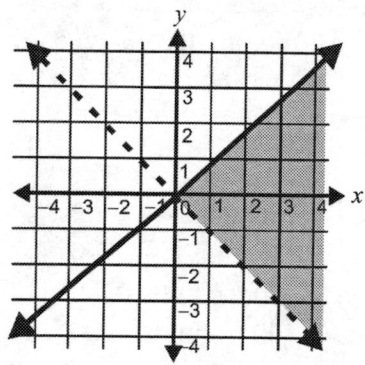

D.

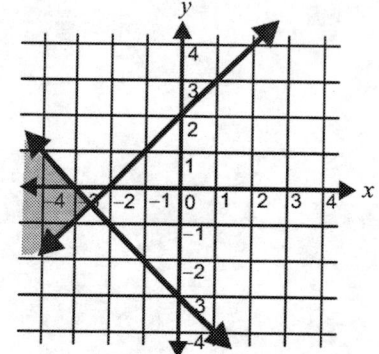

11. Which inequalities are represented by the graph?

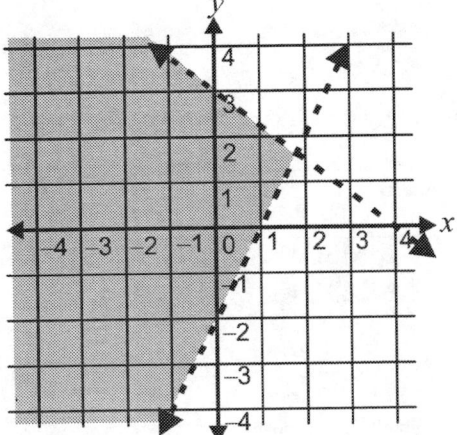

A. $2x + 2y \geq -4$
   $3y < 2x + 6$

B. $7x + 7y \leq 21$
   $8x < 6y - 24$

C. $9x + 12y < 36$
   $34x - 17y < 34$

D. $-11x - 22y \geq 44$
   $-4x + 2y \leq 8$

12. Two numbers have a sum of 210 and a difference of 30. What are the two numbers?

A. 140 and 170
B. 170 and 40
C. 150 and 60
D. 120 and 90

13. Lucy is getting snacks for her party tonight. She wants to keep the price under thirty dollars. She can buy three 12 packs of soda and 2 bags of chips for $27.33 or she can buy two 12 packs of soda and three bags of chips for $25.27. How much is one bag of chips?

A. $4.23
B. $2.62
C. $6.29
D. $1.60

14. Which inequalities are represented by the graph?

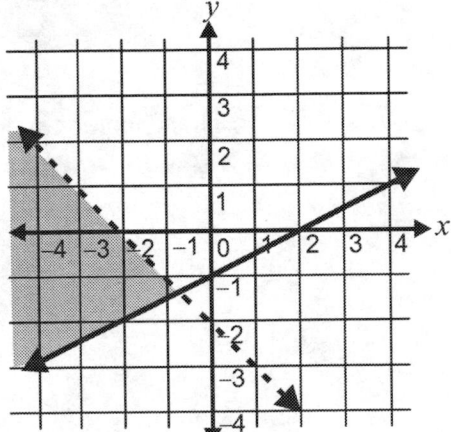

A. $-x \geq -2y - 2$
   $-2x - 2y > 4$

B. $2x - 2y \leq 4$
   $3x + 3y \leq -9$

C. $-3x \leq 6 + 2y$
   $y \geq -x - 2$

D. $3x + 4y \geq 12$
   $y > -3x + 2$

15. The sum of the digits of a two-digit number is seven. When the digits are reversed, the number is increased by 27. What is the number?

A. 25
B. 34
C. 16
D. 27

16. Adam's Concession stand sells 300 nachos and 172 bags of popcorn for a total sale of $1,308 on Friday. The following Friday, they sell 226 nachos and 226 bags of popcorn for a total sale of $1,130. How much is each order of nachos and each bag of popcorn?

A. $3.50/bag of popcorn, $1.50/nachos
B. $1.39/nachos, $3.94/bag of popcorn
C. $1.39/bag of popcorn, $3.94/nachos
D. $3.50/nachos, $1.50/bag of popcorn

Chapter 3
Circles

This chapter covers the following Georgia Performance Standards:

MM3G	Geometry	MM3G1a, MM3G1b, MM3G1c
		MM3G1d, MM3G1e

A **circle** is the set of all points (x, y) in a plane that are a distance r from point (a, b). The distance r is called the **radius** of the circle, and the point (a, b) is called the **center** of the circle. A circle is a specific type of conic section, explained in detail in the next chapter.

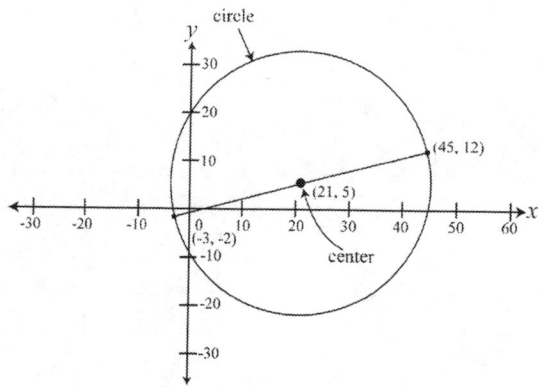

3.1   Finding Equations of Circles

A circle with a radius, r, and a center, (a, b), that is graphed on a coordinate grid can be defined by the equation $(x - a)^2 + (y - b)^2 = r^2$. If the center of the circle is at the origin, $(0, 0)$, then the equation can be simplified to $x^2 + y^2 = r^2$.

Example 1:   Find the equation of the circle with the center: at $(-1, 4)$ and a radius of 15.

Step 1:   The center is $(-1, 4)$, so $a = -1$ and $b = 4$. The first part of the equation is $(x + 1)^2 + (y - 4)^2$.

Step 2:   The radius is 15, so $r^2 = 15^2$ or 225.

The equation of the circle is $(x + 1)^2 + (y - 4)^2 = 225$.

Find the equations of the following circles.

1. Center: $(-2, 10)$; Radius: 13

2. Center: $(12, -1)$; Radius: 21

3. Center: $(-8, -9)$; Radius: 7

4. Center: $(17, 3)$; Radius: 11

5. Center: $(2, -4)$; Radius: 3

6. Center: $(20, -5)$; Radius: 15

7. Find the equation of the following circle where C is the center and A is on the circle.

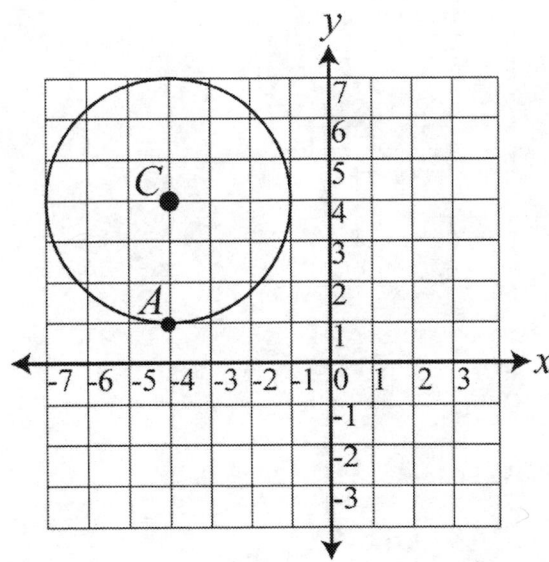

8. A circle has a center at the point $(2, -2)$ and a circumference of 8π units. Find the equation of the circle.

9. A circle has a center at the point $(-5, 3)$ and an area of 31π units^2. Find the equation of the circle.

10. The origin of a coordinate grid lies on a circle with a center at the point $(3, 4)$. Find the equation of the circle.

3.2   Graphing Circles

Example 2:   The equation of a circle is $(x-2)^2 + (y-7)^2 = 7^2$, or $(x-2)^2 + (y-5)^2 = 49$.

The center of the circle is at $(2, 7)$, and the radius is 7.

Step 1:   Plot the point $(2, 7)$, then move up, down, left and right by a distance of the radius. This will generate four points that are on the circle.

$(2, 7)$ up 7 units: $(2, 14)$

$(2, 7)$ down 7 units: $(2, 0)$

$(2, 7)$ left 7 units: $(-5, 7)$

$(2, 7)$ right 7 units: $(9, 7)$

Step 2:   Plot these four points, and then draw a circle that includes all four points.

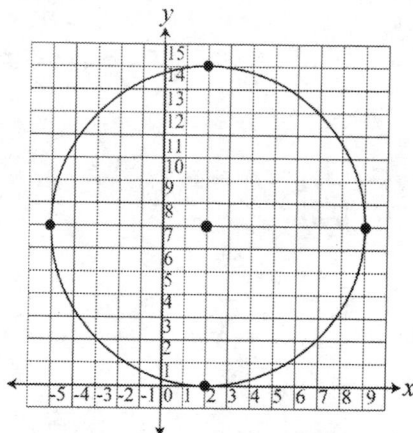

Graph the following circles.

1. $x^2 + y^2 = 49$

2. $(x+2)^2 + y^2 = 81$

3. $x^2 + (y+5)^2 = 4$

4. $(x-2)^2 + (y+3)^2 = 77$

3.3   Finding the Equation of a Line Tangent to a Circle

A line is tangent to a circle at a given point if the line touches the circle only at that point. The tangent line is perpendicular to the line that contains both the radius and the point of tangency.

Example 3:   Find the equation of a tangent line to the circle, $x^2 + y^2 = 52$, at the point $(-4, 6)$.

Step 1:   Find the slope of the line that contains the center, $(0, 0)$ and the point of tangency, $(-4, 6)$.

$$\frac{6-0}{-4-0} = \frac{6}{-4} = -\frac{3}{2}$$

Since the tangent line is perpendicular to the radius at that point, the slope of the tangent line would be $(-1)\left(-\frac{2}{3}\right)$, or $\frac{2}{3}$.

Step 2:   Next, use the point-slope formula to calculate the equation of the tangent line. The slope is $\frac{2}{3}$, and one point on the line is $(-4, 6)$. Plug these values into the point-slope formula, and solve for y.

$$(y - y_1) = m(x - x_1)$$

$$(y - 6) = \tfrac{2}{3}(x + 4)$$

$$y - 6 = \tfrac{2}{3}x + \tfrac{8}{3}$$

$$y = \tfrac{2}{3}x + \tfrac{8}{3} + \tfrac{18}{3}$$

$$y = \tfrac{2}{3}x + \tfrac{26}{3}$$

Find the equation of the tangent line to the circle at the given point.

1. $x^2 + y^2 = 25$, at the point $(3, 4)$.

2. $(x - 1)^2 + y^2 = 25$, at the point $(4, 4)$.

3. $x^2 + (y - 1)^2 = 25$, at the point $(3, 5)$.

4. $(x - 1)^2 + (y - 1)^2 = 25$, at the point $(4, 5)$.

3.4   Solving a System of Equations Involving a Circle and a Line

A circle and a line may not intersect, may intersect at a single point of tangency, or may intersect at two points. Refer to the following three diagrams.

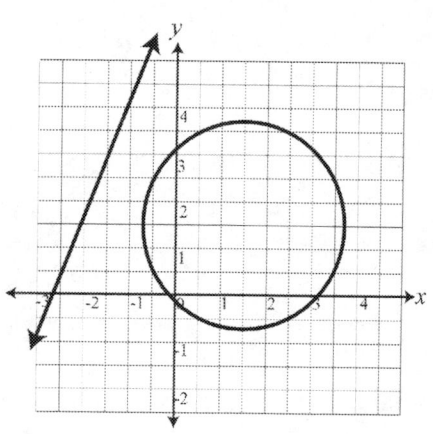 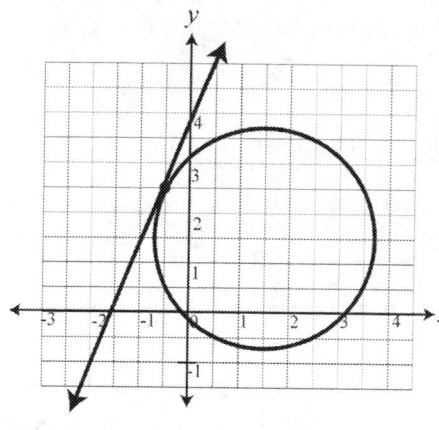 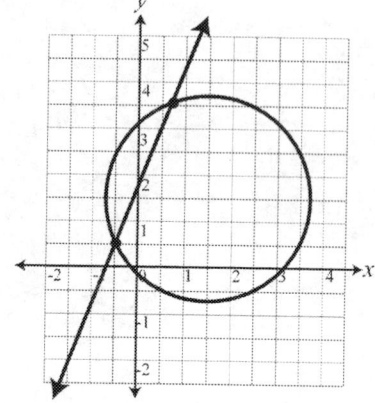

Use the following steps to find where a circle and line intersect, if at all.

Example 4:   If the equation of a circle is $x^2 + y^2 = 16$ and the equation of a line is $y = x - 2$, find the point(s) of intersection, if any.

Step 1:   Substitute $x - 2$ for y in the equation of the circle.
$x^2 + (x - 2)^2 = 16$

Step 2:   Expand: $(x - 2)^2$
$x^2 + x^2 - 4x + 4 = 16$

Step 3:   Set the equation equal to 0, and collect like terms.
$x^2 + x^2 - 4x + 4 - (16) = 16 - (16)$
$2x^2 - 4x - 12 = 0$

Step 4:   Factor by completing the square.

$2(x^2 - 2x - 6) = 0$ Factor out 2.

$\dfrac{2(x^2 - 2x - 6)}{2} = \dfrac{0}{2}$ Divide both sides by 2.

$x^2 - 2x - 6 + 6 = 0 + 6$ Add 6 to both sides.

$x - 2x + \underline{} = 6 + \underline{}$ Complete the square.

$x - 2x + 1 = 6 + 1$ Factor the left side.

$(x - 1)^2 = 7$ Take the square root of both sides.

$x - 1 + 1 = 1 \pm \sqrt{7}$ Add 1 to both sides.

$x = 1 \pm \sqrt{7}$

Step 5:   Substitute the x-values into the equation $y = x - 2$ to find the corresponding y-values.

$$y = (1 + \sqrt{7}) - 2 \qquad y = -1 + \sqrt{7}, \text{ so } (1 + \sqrt{7}, -1 + \sqrt{7}) \text{ or } (3.6, 1.6)$$

$$y = (1 - \sqrt{7}) - 2 \qquad y = -1 - \sqrt{7}, \text{ so } (1 - \sqrt{7}, -1 - \sqrt{7}) \text{ or } (-1.6, -3.6)$$

Step 6:   Interpret your answer.

The line, $y = x - 2$, intersects the circle, $x^2 + y^2 = 16$, at the points $(3.6, 1.6)$ and $(-1.6, -3.6)$.

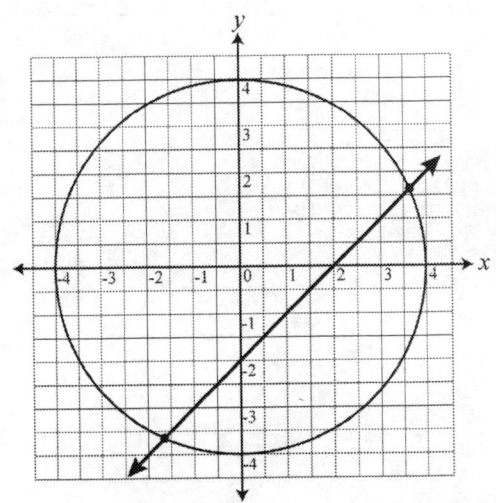

Find the point(s) of intersection, if any, of the following circles and lines.

1. $(x - 7)^2 + (y - 3)^2 = 40 \qquad y = -3x + 4$

2. $(x - 7)^2 + (y - 3)^2 = 40 \qquad y = x$

3. $(x - 7)^2 + (y - 3)^2 = 40 \qquad y = -\frac{7}{4}x - \frac{3}{2}$

4. $(x + 2)^2 + (y)^2 = 10 \qquad y = -x$

5. $(x - 1)^2 + (y)^2 = 16 \qquad y = x + 5$

6. $(x + 2)^2 + (y)^2 = 25 \qquad y = 5$

3.5   Solving a System of Equations Involving Two Circles

Two circles may not intersect, may intersect at a single point of tangency, may intersect at two points, or may be the same circle and intersect at an infinite number of points. Refer to the following four diagrams.

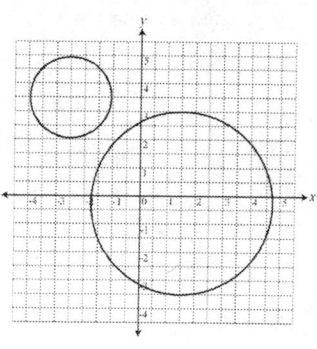
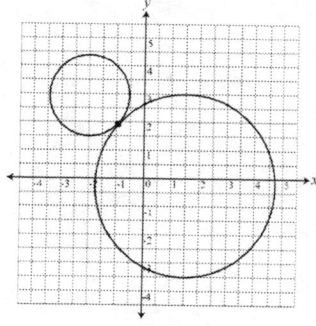

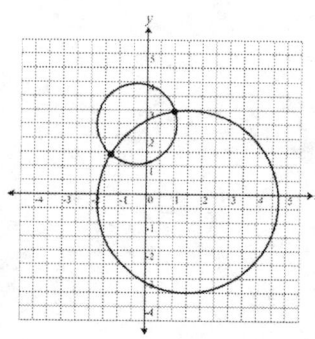
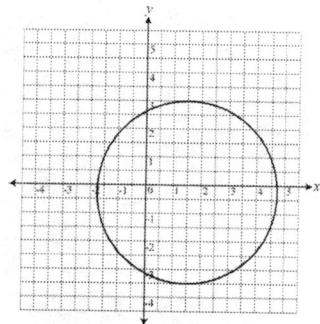

Example 5:   If the equation of Circle A is $(x+1)^2 + (y+1)^2 = 2$, and the equation of circle B is $(x+2)^2 + (y+3)^2 = 1$, find the point(s) of intersection, if any.

Step 1:   Put both equations in the expanded form $Ax^2 + By^2 + Cx + Dy + E = 0$.

Equation of Circle A:
$$(x+1)^2 + (y+1)^2 = 2$$
$$x^2 + 2x + 1 + y^2 + 2y + 1 = 2$$
$$x^2 + y^2 + 2x + 2y = 0$$

Equation of Circle B:
$$(x+2)^2 + (y+3)^2 = 1$$
$$x^2 + 4x + 4 + y^2 + 6y + 9 = 1$$
$$x^2 + y^2 + 4x + 6y + 12 = 0$$

Step 2:   Subtract the equation for Circle B from the equation for Circle A.

$$x^2 + y^2 + 2x + 2y = 0$$
$$- (x^2 + y^2 + 4x + 6y + 12 = 0)$$
$$\overline{\qquad -2x - 4y - 12 = 0}$$

Step 3:   Solve for x.

$$-2x - 4y - 12 = 0$$
$$x = -2y - 6$$

Step 4: Plug $x = -2y - 6$ into the equation for Circle A.

$$(x+1)^2 + (y+1)^2 = 2$$

$$(-2y - 6 + 1)^2 + (y+1)^2 = 2$$

Step 5: Solve for y.

$$(-2y - 6 + 1)^2 + (y+1)^2 = 2$$

$$4y^2 + 20y + 25 + y^2 + 2y + 1 = 2$$

$$5y^2 + 22y + 24 = 0$$

$$(5y + 12)(y + 2) = 0$$

$$y = -2.4 \text{ or } -2$$

Step 6: Substitute these values of y back into the equation for x. $\quad x = -2y - 6$

$$x = -2(-2.4) - 6 \qquad x = -2(-2) - 6$$
$$x = -1.2 \qquad\qquad x = -2$$

Step 7: Interpret your answers.

Circles A and B intersect each other at the points $(-1.2, -2.4)$ and $(-2, -2)$.

Find the points of intersection, if any, of the following circles.

1. $(x+5)^2 + (y-4)^2 = 4$ and $(x-1)^2 + (y-1)^2 = 25$

2. $(x+5)^2 + (y-4)^2 = 4$ and $(x-1)^2 + (y-4)^2 = 16$

3. $(x+5)^2 + (y-4)^2 = 4$ and $(x-5)^2 + (y-1)^2 = 25$

4. $(x+5)^2 + (y-4)^2 = 4$ and $4(x+5)^2 + 4(y-4)^2 = 16$

Chapter 3 Review

Find the equations of the following circles.

1. Center: $(2, -12)$; Radius: 8

2. A circle has a center at the point $(-4, -2)$ and a circumference of 10π units.

3. A circle has a center at the point $(5, 7)$ and an area of 51π units^2.

4. The origin of a coordinate grid lies on a circle with a center at the point $(-3, -2)$.

Graph the following circle.

5. $(x + 2)^2 + (y - 7)^2 = 121$

Find the equation of the tangent line to the circle at the given point.

6. $(x + 1)^2 + (y - 6)^2 = 117$, at the point $(5, -3)$

7. If the equation of a circle is $x^2 + y^2 = 4$ and the equation of a line is $y = \frac{1}{2}x + 3$, find the point(s) of intersection, if any.

Solve the following problem concerning systems equations involving two circles.

8. If the equation of circle A is $x^2 + y^2 = 36$, and the equation of circle B is $(x - 10)^2 + y^2 = 25$, find the point(s) of intersection, if any.

Chapter 3 Test

1. Find the equation of a circle with center $(15, 1)$ and radius 6.

   A. $(x - 15)^2 + (y - 1)^2 = 36$

   B. $(x - 15)^2 + (y - 1)^2 = 6$

   C. $(x + 15)^2 + (y + 1)^2 = 36$

   D. $(x - 15) + (y - 1) = 6$

2. A circle has a center at the point $(5, 2)$ and a circumference of 6π units. Find the equation of the circle.

   A. $(x + 5)^2 + (y + 1)^2 = 9$

   B. $(x - 5)^2 + (y - 2)^2 = 3$

   C. $(x - 5)^2 + (y - 2)^2 = 9$

   D. $(x - 5) + (y - 1) = 9$

3. A circle has a center at the point $(5, 0)$ and an area of 3π units^2. Find the equation of the circle.

   A. $x + 5^2 + y^2 = 3$

   B. $(x - 5)^2 + y^2 = \sqrt{3}$

   C. $(x - 5)^2 + y^2 = 3$

   D. $x^2 + (y - 5)^2 = \sqrt{3}$

4. The origin of a coordinate grid lies on a circle with a center at the point $(5, -2)$. Find the equation of the circle.

   A. $(x - 5)^2 + (y + 2)^2 = 29$

   B. $(x - 5)^2 + (y + 2)^2 = \sqrt{29}$

   C. $(x - 2)^2 + (y + 5)^2 = 29$

   D. $(x + 5)^2 + (y - 2)^2 = \sqrt{29}$

5. Which of the following is the graph of $x^2 + y^2 = 256$?

   A.

   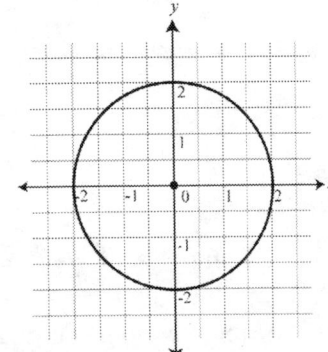

   B.

   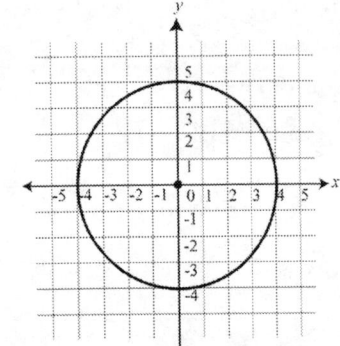

   C.

   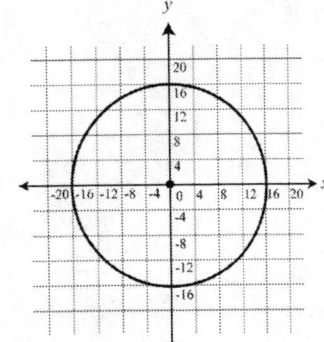

   D.

   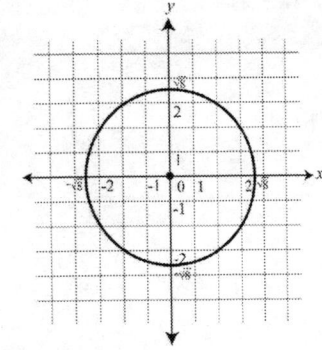

6. Find the equation of the line tangent to the graph of the circle with the equation $(x - 7)^2 + (y - 7)^2 = 157$ at the point $(-4, 1)$.

  A. $y = \frac{6}{11}x - \frac{25}{3}$

  B. $y = -11x - 25$

  C. $y = \frac{11}{6}x + \frac{25}{3}$

  D. $y = -\frac{11}{6}x - \frac{19}{3}$

7. If the equation of a circle is $x^2 + y^2 = 25$ and the equation of a line is $y = -x - 5$, find the point(s) of intersection, if any.

  A. no intersections
  B. $(\sqrt{5}, \sqrt{5})$
  C. $(0, -5), (-5, 0)$
  D. $(12, -17)$, and $(-12, 17)$

8. If the equation of circle A is $x^2 + y^2 = 36$, and the equation of circle B is $(x - 6)^2 + (y - 6)^2 = 36$, find the point(s) of intersection, if any.

  A. $(6, 0)$ and $(0, 6)$
  B. $(0, 6)$
  C. $(0, 0)$
  D. no intersections

Chapter 4
Conics

This chapter covers the following Georgia Performance Standards:

| MM3G | Geometry | MM3G2a, MM3G2b, MM3G2c |

A **conic**, or **conic section**, is defined as a curve that is formed when a plane intersects a double cone. Conics can be divided into four general categories: circles (covered in the previous chapter), parabolas, ellipses, and hyperbolas. All conic sections have equations of the form $Ax^2 + Bxy + Cy^2 + Dx + Ey + F = 0$.

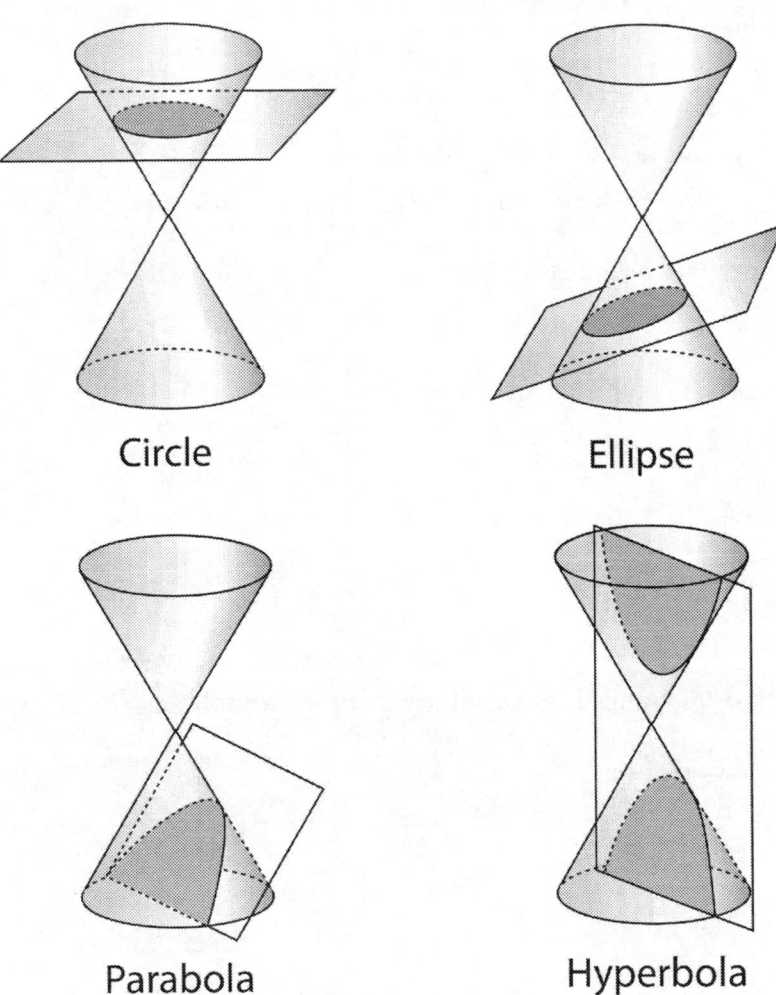

Circle

Ellipse

Parabola

Hyperbola

4.1   Converting Equations of Conics by Completing the Square

Sometimes, it is necessary to "complete the square" in order to more easily graph a conic or identify its characteristics. After completing the square, the equation will be in standard form. We will learn more about the following standard forms of equations in the next section.

Conic Section	Standard Form of its Equation
Parabola	$y = a(x - h)^2 + k$, where (h, k) is the vertex of the parabola
Circle	$(x - h)^2 + (y - k)^2 = r^2$, where r is the radius and (h, k) is the center
Ellipse	$\dfrac{(x - h)^2}{a^2} + \dfrac{(y - k)^2}{b^2} = 1$, where (h, k) is the center
Hyperbola	$\dfrac{(x - h)^2}{a^2} - \dfrac{(y - k)^2}{b^2} = 1$, where (h, k) is the center

Example 1:   Complete the square for the following equation: $9x^2 - 36x + 4y^2 = 0$

Step 1:   Factor 9 from the two x terms.

$$9(x^2 - 4x) + 4y^2 = 0$$

Step 2:   Remember to add 9 times the amount.

$$9(x^2 - 4x + ?) + 4y^2 = 0 + 9(?)$$

Step 3:   Take half of -4, then square it.

$$9(x^2 - 4x + 4) + 4y^2 = 9(4)$$

Step 4:   Factor the completed-square expression.

$$9(x - 2)^2 + 4y^2 = 36$$

Step 5:   Divide both sides by 36 so that the equation will equal 1.

$$\frac{(x - 2)^2}{4} + \frac{y^2}{9} = 1$$

Convert these equations of conics into standard form by completing the square.

1. $\frac{1}{2}x^2 - 10x - y + 56 = 0$

2. $2x^2 + 9y^2 - 126y + 423 = 0$

3. $36x^2 - 25y^2 + 72x + 150y - 1089 = 0$

4.2 Graphing Conic Sections

A **parabola** is an open curve that is the graph of a two-variable quadratic equation in the form $y = ax^2 + bx + c$, where a, b, and c are constants. A parabola can open up, down, left, or right. The quadratic equation can also be written in the form $y = a(x - h)^2 + k$, where the point (h, k) is the **vertex** of the parabola, and the **line of symmetry** is $x = h$. If $a < 0$, the parabola will point either downward or to the left. Every parabola has a **focus** and a **directrix**, such that the distance from any point on the parabola to the focus is the same as the distance from that same point to the directrix. The directrix is perpendicular to the line of symmetry, and the vertex is halfway between the focus and the directrix.

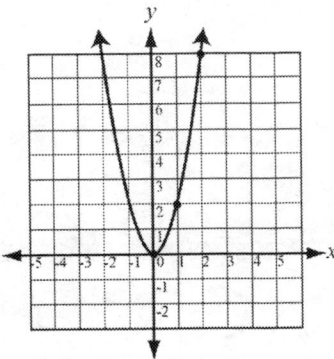

The standard form of the equation of a parabola with its vertex at $(0, 0)$, is $x^2 = 4py$, focus at $(0, p)$, directrix at $y = -p$, vertical axis of symmetry at $x = 0$ or $y^2 = 4px$, focus at $(p, 0)$, directrix at $x = -p$, horizontal axis of symmetry at $y = 0$.

Graph the following parabolas and identify the focus, axis of symmetry, and directrix.

1. $y = 2x^2$

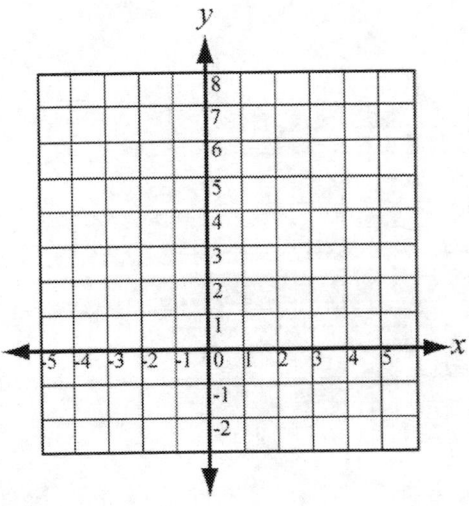

2. $x = \frac{1}{6}y^2$

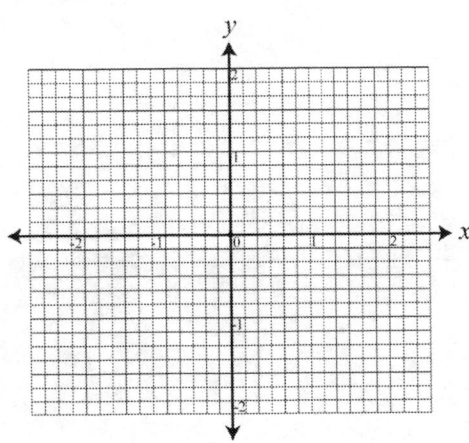

An **ellipse** is an oval-shaped, closed curve that has a major axis (endpoints are **vertices**) and a minor axis (endpoints are co-vertices). An ellipse is the set of all points that are the sum of the distances between the points and two fixed points called foci. The **foci** are always on the **major axis** and are located at the points $(0, \pm c)$ when the major axis is vertical or $(\pm c, 0)$ when the major axis is horizontal. The values a and b determine whether an ellipse is either vertical or horizontal. The larger of the two values is always assigned as the a variable. If the larger value, a, is under the x variable, the ellipse is horizontal. If the a value is under the y variable, the ellipse will be vertical. Use $c^2 = a^2 - b^2$ to find c. When the major axis and the minor axis are the same length, the result is a circle. An ellipse with a major axis length of $2a$, a minor axis length of $2b$, and a center of (h, k) can be defined by the equation $\dfrac{(x - h)^2}{a^2} + \dfrac{(y - k)^2}{b^2} = 1$.

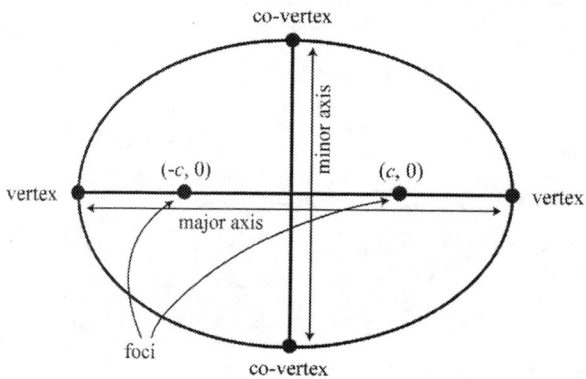

Graph the following ellipses. Identify the center, vertices, co-vertices, and foci.

1. $\dfrac{(x - 5)^2}{4} + \dfrac{(y - 11)^2}{1} = 1$

2. $\dfrac{(x + 3)^2}{16} + \dfrac{(y - 4)^2}{36} = 1$

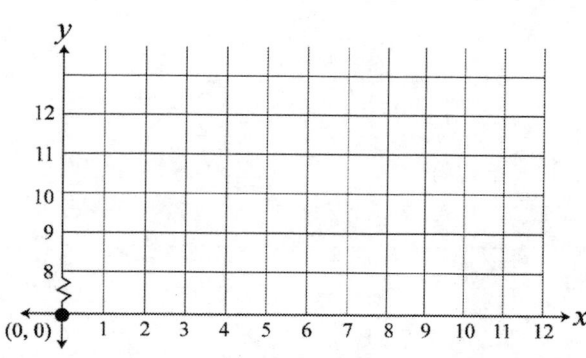

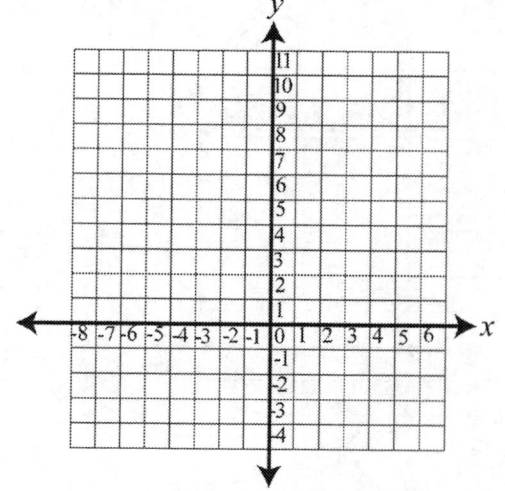

A **hyperbola** is two open curves separated by the distance of the major axis. These two open curves have **asymptotes**, imaginary lines that the graph approaches but never intersect. A hyperbola with a **major axis** length of $2a$, a **minor axis** length of $2b$, and a **center** of (h, k) can be defined by the equation $\frac{(x-h)^2}{a^2} - \frac{(y-k)^2}{b^2} = 1$. If the equation starts with the x variable, the hyperbola will be **horizontal** and the equations of the **asymptotes** are $y = \pm \frac{b}{a}x$; **vertical** hyperbolas start with the y *variable* and have asymptotes at $y = \pm \frac{a}{b}x$. Regardless, the a is always first in the main equation. Use $c^2 = a^2 + b^2$ to find c, and notice that this formula is slightly different from the corresponding parabola formula. Below is a horizontal hyperbola.

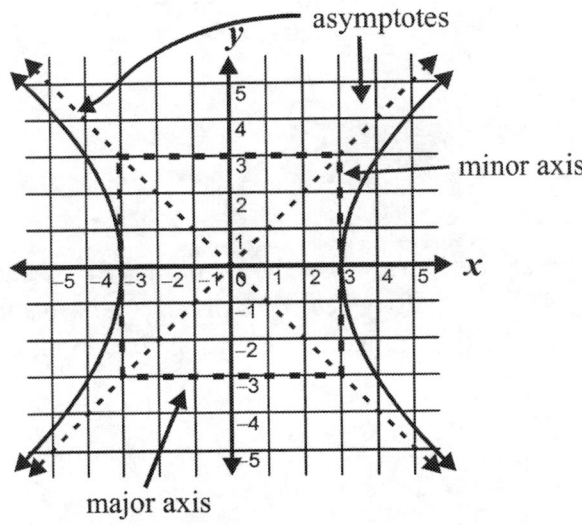

Graph the following hyperbolas and identify the center, vertices, asymptotes, and foci.

1. $\dfrac{(x-5)^2}{4} - \dfrac{(y-11)^2}{1} = 1$

2. $\dfrac{(y-4)^2}{36} - \dfrac{(x+3)^2}{16} = 1$

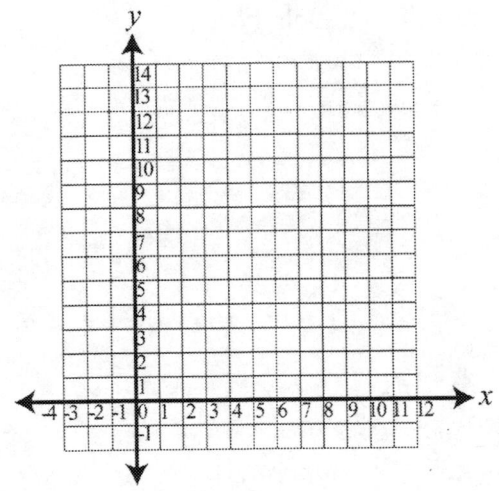

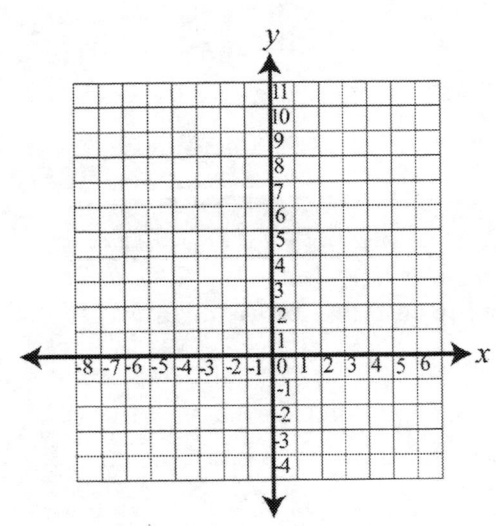

4.3   Writing Equations of Parabolas

Example 2:   The parabola shown below has its vertex at the point $(15, 8)$. Write its equation in the form $y = ax^2 + bx + c$, where a, b, and c are constants.

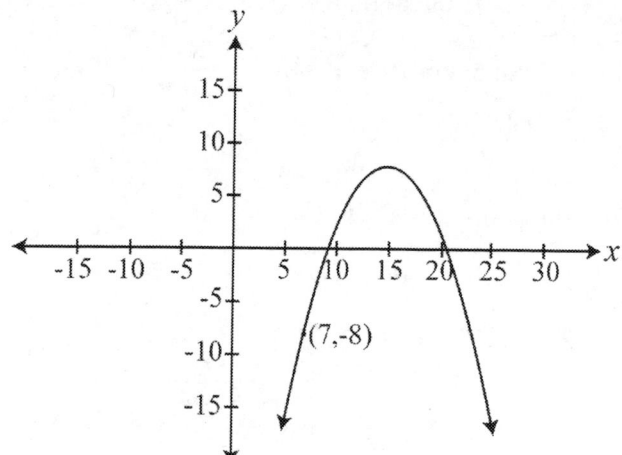

Step 1:   Since the parabola's vertex is at the point $(15, 8)$, its equation can be written as $y = a(x - 15)^2 + 8$. However, it's still necessary to solve for a. To do this, the point $(7, -8)$ should be substituted into the equation, and a should be solved for as follows:

$$y = a(x - 15)^2 + 8$$
$$-8 = a(7 - 15)^2 + 8$$
$$-8 = a(-8)^2 + 8$$
$$-8 = 64a + 8$$
$$-8 - 8 = 64a + 8 - 8$$
$$-16 = 64a$$
$$a = -\frac{16}{64}, \text{ or } -\frac{1}{4}$$

Step 2:   Because the value of a is $-\frac{1}{4}$, the equation of the parabola becomes $y = -\frac{1}{4}(x - 15)^2 + 8$. Transform the equation into the form $y = ax^2 + bx + c$.

$$y = -\frac{1}{4}(x - 15)^2 + 8$$
$$y = -\frac{1}{4}(x^2 - 30x + 225) + 8$$
$$y = -\frac{1}{4}x^2 + \frac{15}{2}x - \frac{225}{4} + 8$$
$$y = -\frac{1}{4}x^2 + \frac{15}{2}x - \frac{193}{4}$$

Determine the equation of the parabola in the form $y = ax^2 + bx + c$ given the vertex and a point.

1. Vertex: $(-7, 1)$; Point: $(1, 129)$

2. Vertex: $(5, 2)$; Point: $(0, 17)$

3. Vertex: $(3, -3)$; Point: $(2, 3)$

4. Vertex: $(-1, -2)$; Point: $(-2, 3)$

4.4   Writing Equations of Ellipses

Example 3:   The ellipse shown below has its center at the point $(3, 7)$. Determine its equation.

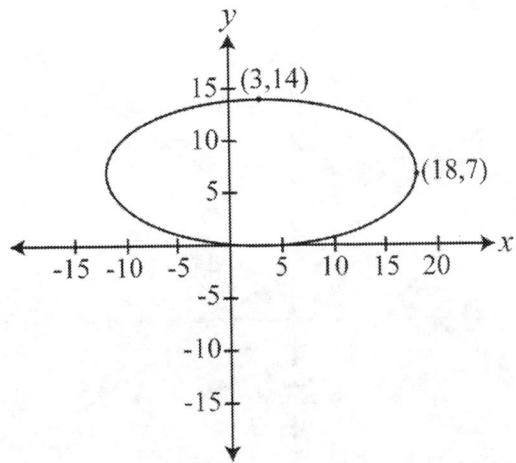

Step 1:   The center of the ellipse is at the midpoint of the major axis and also at the midpoint of the minor axis. The center is at the point $(3, 7)$. It is given that the point $(18, 7)$ is on the major axis, and that the point $(3, 14)$ is on the minor axis.

Step 2:   Because the center is at the midpoint of these axes, the distance from the center, or point $(3, 7)$, to point $(18, 7)$ is half the length of the major axis, while the distance from point $(3, 7)$ to point $(3, 14)$ is half the length of the minor axis.

Step 3:   Therefore the length of the major axis is 30 units, and the length of the minor axis is 14 units, so the equation of the ellipse is $\dfrac{(x - 3)^2}{225} + \dfrac{(y - 7)^2}{49} = 1$.

Find the equation of the ellipse that contains the following:

1. Center: $(2, 4)$; Points: $(2, 8)$, $(12, 4)$

2. Center $(-3, 4)$; Points: $(1, 4)$, $(-3, 10)$

3. Center $(19, 2)$; Points: $(19, -6)$, $(2, 2)$

4. Center: $(-1, -8)$; Points: $(-1, 9)$, $(12, -8)$

4.5   Writing Equations of Hyperbolas

Example 4:   The hyperbola shown below has its center at the point $(-2, 6)$. Determine its equation.

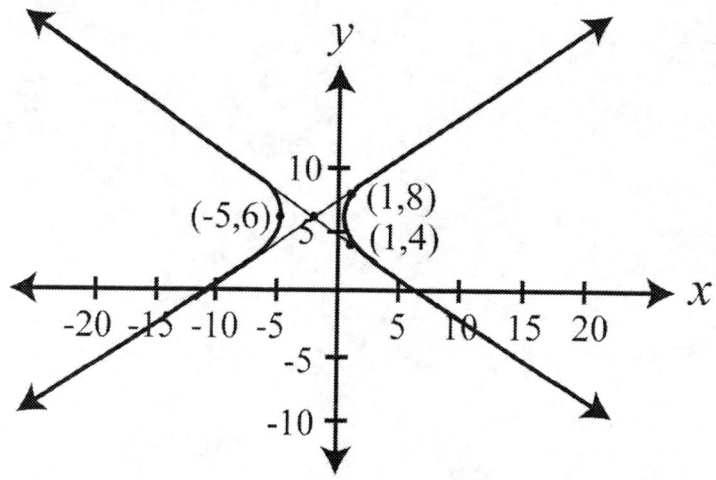

Step 1:   The center of the hyperbola is half the distance between the two curves, so the length of the major axis is twice the distance between the center and one of the curves. The length of the major axis is 6, or 2 times $-2 - (-5)$. Therefore, $a = 3$.

Step 2:   The length of the minor axis is twice the distance between the points where the line tangent to the vertex of one of the curves intersects the asymptotes. It's clear from the graph that the distance between the points where the line $x = 1$ intersects the asymptotes is $8 - 4$ or 4, so half this distance is 2. Therefore, the length of the minor axis is 4 and $b = 2$.

The equation of the hyperbola is $\dfrac{(x - h)^2}{a^2} - \dfrac{(y - k)^2}{b^2} = 1$.

Substitute known values into the equation.

$$\dfrac{(x + 2)^2}{9} - \dfrac{(y - 6)^2}{4} = 1.$$

Determine the equation of the hyperbola that satisfies each of the following sets of conditions.

1. Center: $(11, -4)$; Point on Curve: $(7, -4)$; Points on Asymptotes: $(15, 1)$, $(15, -9)$

2. Center: $(-5, 3)$; Point on Curve: $(1, 3)$; Points on Asymptotes: $(-11, 6)$, $(-11, 0)$

3. Center: $(2, 9)$; Point on Curve: $(9, 9)$; Points on Asymptotes: $(9, 0)$, $(9, 18)$

Chapter 4 Review

1. Convert the following equation to standard form by completing the square.
   $3x^2 - 12x + 29 - y = 0$

2. Graph the equation $y = -x^2 + 4$.

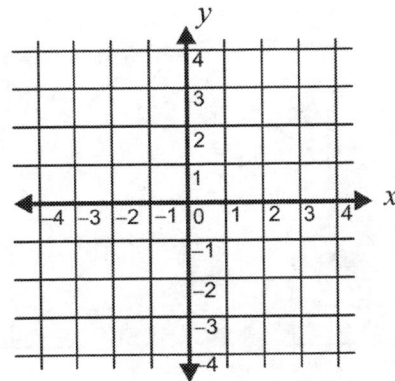

3. Find the vertex of $f(x) = 2x^2 - 8x + 15$.

4. Graph the equation $y = -\frac{1}{2}x^2 + 1$.

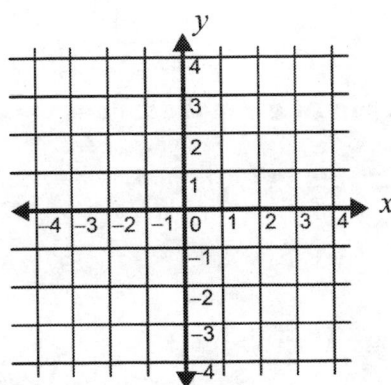

5. Graph $y = \frac{1}{5}x^2$ and label the focus, axis of symmetry, and the directrix.

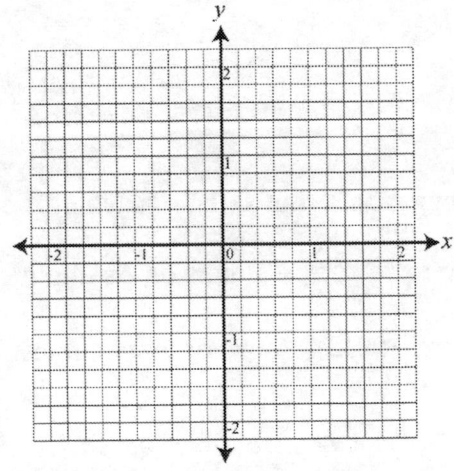

For problems 6 and 7, graph the equations, then find the center, vertices, co-vertices (only for problem 6), and foci.

6. $\dfrac{(x-19)^2}{289} + \dfrac{(y-2)^2}{64} = 1$

7. $\dfrac{(x-19)^2}{289} - \dfrac{(y-2)^2}{64} = 1$

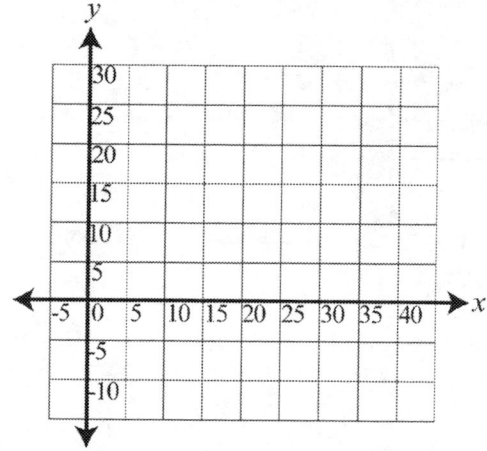

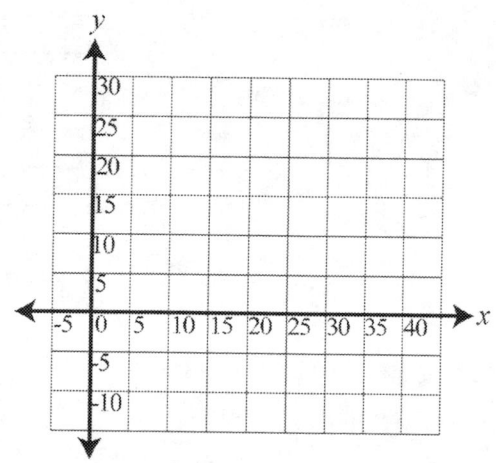

Find the equation of the ellipse given the following center and points.

8. Center: $(5, 11)$; Points: $(5, 12)$, $(3, 11)$

9. Center: $(9, 8)$; Points: $(-10, 8)$, $(9, 1)$

10. Center: $(-2, 17)$; Points: $(-2, 20), (2, 17)$

Find the equation of the parabola in the form $y = ax^2 + bx + c$ given the vertex and a point on the parabola.

11. Vertex: $(-9, 3)$; Point: $(-8, 1)$

12. Vertex: $(4, 32)$; Point: $(3, 29)$

Determine the equation of the hyperbola that satisfies each of the following sets of conditions.

13. Center: $(5, -5)$; Point on Curve: $(-1, -5)$; Points on Asymptotes: $(11, -12)$, $(11, 2)$

14. Center: $(-3, 2)$; Point on Curve: $(20, 2)$; Points on Asymptotes: $(20, -18)$, $(20, 22)$

Chapter 4 Test

1. The equation of a hyperbola is
$$10x^2 - 8y^2 - 40x - 32y - 72 = 0.$$
Which of the following is the equation in standard form?

A. $\dfrac{(x-2)^2}{8} + \dfrac{(y+2)^2}{10} = 1$

B. $\dfrac{(x-2)^2}{8} - \dfrac{(y+2)^2}{10} = 1$

C. $\dfrac{(y-2)^2}{8} - \dfrac{(x+2)^2}{10} = 1$

D. $\dfrac{(x-2)^2}{100} - \dfrac{(y+2)^2}{64} = 1$

2. What is the equation for the axis of symmetry of the following parabola?
$$y = 3(x-4)^2 + 2$$

A. $y = 4$

B. $x = 4$

C. $y = 2$

D. $x = -2$

3. Which of the following is the ordered pair for one of the foci of the ellipse?
$$\frac{(x+1)^2}{169} + \frac{(y+8)^2}{36} = 1$$

A. $(-1, -8)$

B. $(-1 - \sqrt{133}, -8)$

C. $(-1 - \sqrt{133}, 8)$

D. $(1 + \sqrt{133}, 8)$

4. In the hyperbola $\dfrac{(x+1)^2}{169} - \dfrac{(y+8)^2}{289} = 1$, what are the coordinates of the foci?

A. $(-22.4, -8), (20.4, -8)$

B. $(-1, -8), (1, 8)$

C. $(-14, -8), (12, 8)$

D. $(-8, -14), (8, 12)$

5. The equation of a hyperbola is
$$\frac{x^2}{121} - \frac{y^2}{144} = 1.$$
Which of the following is an equation of one of its asymptotes?

A. $y = -\dfrac{11}{12}x$

B. $y = -\dfrac{144}{121}x$

C. $y = -\dfrac{12}{11}x$

D. $y = \dfrac{121}{144}x$

6. Which of the following equations is a parabola with a vertex at $(-10, -3)$ and a point on the parabola is $(-8, 45)$?

A. $y = x^2 + 20x + 97$

B. $y = x^2 + 20x + 100$

C. $y = 12x^2 + 240x + 1197$

D. $y = 12x^2 + 240x + 1203$

7. Which of the following graphs represents $y = 2x^2$?

A.

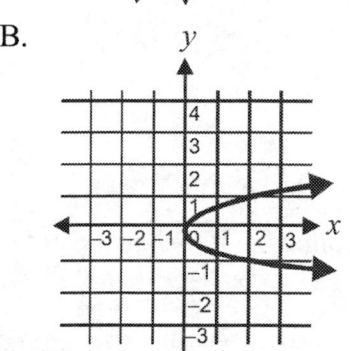

B.

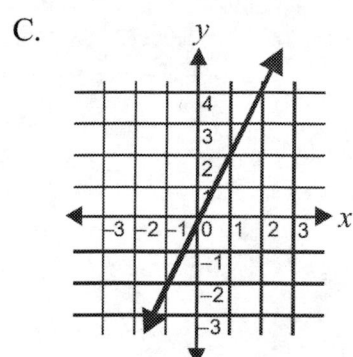

C.

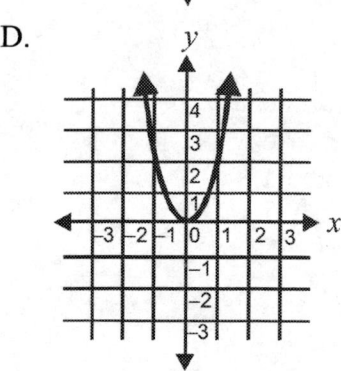

D.

8. Find the equation of the parabola in the form $y = ax^2 + bx + c$ given the vertex and a point on the parabola.

Vertex: $(-9, 3)$; Point: $(-8, 1)$

A. $y = 2(x - 9)^2 - 3$
B. $y = -2(x + 9)^2 + 3$
C. $y = 2x^2 + 36x + 159$
D. $y = -2x^2 - 36x - 159$

9. Find the equation of the following ellipse.

Center: $(20, 6)$; Points: $(10, 6), (20, -2)$

A. $\dfrac{(x - 20)^2}{64} + \dfrac{(y + 6)^2}{100} = 1$

B. $\dfrac{(x - 64)^2}{400} + \dfrac{(y + 100)^2}{36} = 1$

C. $\dfrac{(x - 20)^2}{100} + \dfrac{(y - 6)^2}{64} = 1$

D. $\dfrac{(x - 20)^2}{400} + \dfrac{(y - 6)^2}{36} = 1$

10. The equation of a hyperbola is
$$\frac{(x + 12)^2}{121} - \frac{(y - 10)^2}{144} = 1.$$

Which of the following points does not lie on one of its asymptotes?

A. $(-23, -22)$

B. $(-23, -2)$

C. $(-23, 22)$

D. $(-1, -2)$

Chapter 5
Exponential and Logarithmic Functions

This chapter covers the following Georgia Performance Standards:

MM3A	Algebra	MM3A2a, MM3A2b, MM3A2c, MM3A2d, MM3A2e, MM3A2f, MM3A2g

5.1 Understanding the Properties of nth Roots

The nth root of the number a is written $\sqrt[n]{a}$, and n^{th} roots follow the same properties as square roots. See the following properties.

Basic Properties of Roots

$$\sqrt[n]{ab} = \sqrt[n]{a} \times \sqrt[n]{b} \qquad \sqrt[n]{a^m} = a^{\frac{m}{n}} \qquad \sqrt[n]{\frac{a}{b}} = \frac{\sqrt[n]{a}}{\sqrt[n]{b}}$$

Example 1: What is $\sqrt[5]{-32}$?

Step 1: Determine if the nth root is positive or negative.

To find $\sqrt[5]{-32}$ think of a number that when used 5 times in a multiplication problem with itself produces a product of -32. The number can't be positive, because a positive number multiplied by itself will always produce a product that is positive. This means that $\sqrt[5]{-32}$ must be a negative number.

Step 2: Determine if the nth root is even or odd.

Also, an even number used in a multiplication problem with itself will always produce a product that is even, and an odd number multiplied by itself will always produce a product that is odd. This means that since -32 is an even number, $\sqrt[5]{-32}$ must also be even.

Step 3: Use trial and error.

Since $\sqrt[5]{-32}$ is a negative even number, a good number to test first is -2. For $\sqrt[5]{-32}$ to be equal to -2, $(-2) \times (-2) \times (-2) \times (-2) \times (-2)$ must equal -32, which it does. Therefore, $\sqrt[5]{-32} = -2$.

Step 4: Using the property $\sqrt[n]{a^m} = a^{\frac{m}{n}}$, $\sqrt[5]{-32} = (-32)^{\frac{1}{5}}$.
To solve $\sqrt[5]{-32}$ using a calculator, plug $(-32)\,\hat{}\,(1/5)$ into the calculator.

Find each of the following roots.

1. $\sqrt[3]{343}$

2. $\sqrt[4]{625}$

3. $\sqrt{289}$

4. $\sqrt[3]{-216}$

5. $\sqrt[5]{-1024}$

6. $\sqrt[4]{81}$

Example 2:   Simplify the expression $\sqrt[3]{5} \times \sqrt[3]{25}$.

Step 1:   Use the property $\sqrt[n]{ab} = \sqrt[n]{a} \times \sqrt[n]{b}$.

$$\sqrt[3]{5} \times \sqrt[3]{25} = \sqrt[3]{5 \times 25}$$

Step 2:   Since $25 = 5 \times 5$, then $\sqrt[3]{5 \times 25} = \sqrt[3]{5 \times 5 \times 5}$.

Step 3:   Simplify the radical using the property $\sqrt[n]{a^m} = a^{\frac{m}{n}}$.

$$\sqrt[3]{5 \times 5 \times 5} = \sqrt[3]{5^3} = 5^{\frac{3}{3}} = 5^1 = 5$$

Step 4:   Using the properties $\sqrt[n]{ab} = \sqrt[n]{a} \times \sqrt[n]{b}$ and $\sqrt[n]{a^m} = a^{\frac{m}{n}}$,
$\sqrt[3]{5} \times \sqrt[3]{25} = \sqrt[3]{125} = (25)^{\frac{1}{3}}$.
To solve $\sqrt[3]{5} \times \sqrt[3]{25}$ using a calculator, plug $(125)\,\hat{}\,(1/3)$ into the calculator.

Simplify each of the following expressions.

7. $\sqrt[4]{8} \times \sqrt[4]{2}$

8. $\dfrac{\sqrt[3]{81}}{\sqrt[3]{3}}$

9. $\sqrt{7} \times \sqrt{28}$

10. $\dfrac{\sqrt[5]{400,000}}{\sqrt[5]{4}}$

11. $\sqrt[3]{32} \times \sqrt[3]{16}$

12. $\dfrac{\sqrt[4]{1280}}{\sqrt[4]{5}}$

Example 3:   Solve the equation $\sqrt[7]{4^5} = 4^x$ for x.

Step 1:   Use the property $\sqrt[n]{a^m} = a^{\frac{m}{n}}$.

Using the formula for our equation, we see that $n = 7$ and $m = 5$. Therefore, $\sqrt[7]{4^5} = 4^{\frac{5}{7}}$.

Step 2:   Solve for x.

This means that the value of x is $\frac{5}{7}$.

Solve each of the following equations for x.

13. $\sqrt[9]{6^2} = 6^x$

14. $\sqrt[4]{12^3} = 12^x$

15. $\sqrt{8^5} = 8^x$

16. $\sqrt[6]{2^7} = 2^x$

17. $\sqrt[3]{5^{10}} = 5^x$

18. $\sqrt[11]{7^8} = 7^x$

5.2   Multiplying with Fractional Exponents

To multiply two expressions with the same base, add the exponents together and keep the base the same. Numbers with **fractional exponents** follow the same rules as numbers with whole number exponents.

Example 4: $(4)^{\frac{1}{2}} \times (4)^{\frac{3}{2}} = 4^{\frac{1}{2}+\frac{3}{2}} = 4^{\frac{4}{2}} = 4^2$

Example 5: $2x^{\frac{3}{4}} \times 5x^{\frac{1}{5}} = 10x^{\frac{3}{4}+\frac{1}{5}} = 10x^{\frac{15}{20}+\frac{4}{20}} = 10x^{\frac{19}{20}}$

Notice that only the "x" is raised to a power and not the 2 or the 5.

Example 6: $(2x)^{\frac{2}{3}} = 2^{\frac{2}{3}} \times x^{\frac{2}{3}} = 2^{\frac{2}{3}}x^{\frac{2}{3}}$

Example 7: $(3a^2)^{\frac{1}{4}} = 3^{\frac{1}{4}}a^{\frac{2}{4}} = 3^{\frac{1}{4}}a^{\frac{1}{2}}$

Simplify each of the expressions below.

1. $y^{\frac{1}{3}} \times y^{\frac{2}{3}}$

2. $(2)^{\frac{5}{9}} \times (2)^{\frac{1}{3}}$

3. $(10)^{\frac{4}{7}} \times (10)^{\frac{5}{7}}$

4. $b^{\frac{1}{2}} \times b^{\frac{1}{2}}$

5. $x^{\frac{1}{6}} \times x^{\frac{2}{3}}$

6. $(6)^{\frac{1}{12}} \times (6)^{\frac{3}{4}}$

7. $(3)^{\frac{5}{12}} \times (3)^{\frac{7}{2}}$

8. $(12)^{\frac{13}{25}} \times (12)^{\frac{1}{5}}$

9. $2a^{\frac{7}{9}} \times a^{\frac{4}{9}}$

10. $(8)^{\frac{2}{5}} \times (8)^{\frac{1}{10}}$

11. $4a^{\frac{6}{7}} \times 4a^{\frac{4}{5}}$

12. $3(x)^{\frac{4}{13}} \times 5(x)^{\frac{1}{2}}$

13. $(5)^{\frac{1}{2}}$

14. $(7x)^{\frac{6}{7}}$

15. $(14x^2)^{\frac{2}{9}}$

16. $\left(x^{\frac{5}{7}}\right)^{\frac{1}{5}}$

17. $(x^3)^{\frac{1}{3}}$

18. $(4x^5)^{\frac{4}{25}}$

19. $(23y)^{\frac{1}{2}}$

20. $\left(3x^{\frac{1}{2}}\right)^{\frac{2}{5}}$

21. $(16)^{\frac{1}{2}}$

22. $(11x^2)^{\frac{5}{2}}$

23. $(8)^{\frac{6}{13}}$

24. $\left(4a^{\frac{7}{2}}\right)^{\frac{1}{2}}$

25. $(5t)^{\frac{9}{11}}$

26. $(z^9)^{\frac{1}{9}}$

27. $(17)^{\frac{3}{4}}$

28. $\left(9y^{\frac{1}{2}}\right)^{\frac{1}{3}}$

29. $(8)^{\frac{1}{4}}$

30. $(33x)^{\frac{4}{7}}$

31. $(x^4)^{\frac{3}{8}}$

32. $\left(12x^{\frac{7}{9}}\right)^{\frac{3}{7}}$

33. $(4w)^{\frac{1}{6}}$

34. $(z^2)^{\frac{2}{13}}$

35. $(15)^{\frac{1}{15}}$

36. $\left(5a^{\frac{2}{9}}\right)^{\frac{3}{10}}$

5.3   Dividing with Fractional Exponents

Fractional exponents that have the same base can also be divided.

Example 8: $\dfrac{5^{\frac{3}{4}}}{5^{\frac{1}{2}}}$ This problem means $5^{\frac{3}{4}} \div 5^{\frac{1}{2}}$.

Solution: $\dfrac{5^{\frac{3}{4}}}{5^{\frac{1}{2}}} = 5^{\frac{3}{4}-\frac{1}{2}} = 5^{\frac{3}{4}-\frac{2}{4}} = 5^{\frac{1}{4}}$

A quick way to simplify this problem is to subtract the exponents. **When dividing exponents with the same base, subtract the exponents.**

Simplify the problems below.

1. $\dfrac{x^{\frac{7}{9}}}{x^{\frac{1}{9}}}$

2. $\dfrac{2^{\frac{3}{4}}}{2^{\frac{3}{5}}}$

3. $\dfrac{12^{\frac{6}{7}}}{12^{\frac{1}{2}}}$

4. $\dfrac{a^{\frac{1}{3}}}{a^{\frac{1}{6}}}$

5. $\dfrac{6^{\frac{3}{5}}}{6^{\frac{1}{7}}}$

6. $\dfrac{x^{\frac{13}{25}}}{x^{\frac{2}{5}}}$

7. $\dfrac{y^{\frac{8}{9}}}{y^{\frac{2}{3}}}$

8. $\dfrac{8^{\frac{3}{4}}}{8^{\frac{1}{5}}}$

9. $\dfrac{2^{\frac{14}{15}}}{2^{\frac{4}{15}}}$

10. $\dfrac{7^{\frac{6}{7}}}{7^{\frac{2}{7}}}$

11. $\dfrac{b^{\frac{1}{2}}}{b^{\frac{1}{6}}}$

12. $\dfrac{5^{\frac{9}{10}}}{5^{\frac{2}{5}}}$

13. $\dfrac{4^{\frac{9}{17}}}{4^{\frac{1}{2}}}$

14. $\dfrac{9^{\frac{13}{15}}}{9^{\frac{2}{3}}}$

15. $\dfrac{17^{\frac{2}{5}}}{17^{\frac{1}{7}}}$

16. $\dfrac{3^{\frac{2}{3}}}{3^{\frac{1}{12}}}$

17. $\dfrac{x^{\frac{3}{4}}}{x^{\frac{4}{9}}}$

18. $\dfrac{2^{\frac{3}{2}}}{2^{\frac{9}{10}}}$

19. $\dfrac{16^{\frac{5}{3}}}{16^{\frac{7}{8}}}$

20. $\dfrac{10^{\frac{3}{24}}}{10^{\frac{1}{24}}}$

5.4 More Rational Exponents

Example 9: Solve the equation $7^{\frac{2}{3}} \times 7^{\frac{1}{4}} = 7^x$ for x.

Step 1: Use the property $a^m \cdot a^n = a^{m+n}$.

$$7^{\frac{2}{3}} \times 7^{\frac{1}{4}} = 7^{\frac{8}{12}} \times 7^{\frac{3}{12}} = 7^{\frac{8}{12}+\frac{3}{12}} = 7^{\frac{8+3}{12}} = 7^{\frac{11}{12}}$$

Step 2: Solve for x.

$$7^{\frac{2}{3}} \times 7^{\frac{1}{4}} = 7^{\frac{11}{12}} = 7^x, \text{ so } x = \frac{11}{12}.$$

Solve each of the following equations for x.

1. $5^{14} \times 5^{12} = 5^x$

2. $\dfrac{9^{10}}{9^8} = 9^x$

3. $11^{\frac{2}{5}} \times 11^{\frac{1}{10}} = 11^x$

4. $3^{\frac{1}{2}} \div 3^{\frac{1}{6}} = 3^x$

5. $6^{\frac{3}{8}} \times 6^{\frac{3}{4}} = 6^x$

6. $\dfrac{10^{\frac{1}{3}}}{10^{\frac{2}{9}}} = 10^x$

Example 10: Solve the equation $\left(4^{\frac{5}{8}}\right)^{\frac{1}{2}} = 4^x$ for x.

Step 1: $\left(4^{\frac{5}{8}}\right)^{\frac{1}{2}} = 4^{\frac{5}{8} \times \frac{1}{2}} = 4^{\frac{5 \times 1}{8 \times 2}} = 4^{\frac{5}{16}}$

Step 2: $\left(4^{\frac{5}{8}}\right)^{\frac{1}{2}} = 4^{\frac{5}{16}} = 4^x, \text{ so } x = \frac{5}{16}.$

Solve each of the following equations for x.

7. $\left(2^5\right)^5 = 2^x$

8. $\left(7^{-4}\right)^3 = 7^x$

9. $\left(8^{\frac{1}{7}}\right)^{\frac{5}{6}} = 8^x$

10. $\left(12^4\right)^{\frac{1}{16}} = 12^x$

11. $\left(5^{\frac{5}{11}}\right)^{\frac{1}{5}} = 5^x$

12. $\left(20^{\frac{2}{3}}\right)^{\frac{2}{3}} = 20^x$

Example 11: Simplify the expression $14x^{\frac{1}{4}} + x^{\frac{3}{4}} - 10x^{\frac{1}{4}} + 2x^{\frac{3}{4}}$.

Step 1: Rearrange the terms.

$$14x^{\frac{1}{4}} + x^{\frac{3}{4}} - 10x^{\frac{1}{4}} + 2x^{\frac{3}{4}} = 14x^{\frac{1}{4}} - 10x^{\frac{1}{4}} + x^{\frac{3}{4}} + 2x^{\frac{3}{4}}$$

Step 2: Combine like terms (terms with the same exponent).

$$14x^{\frac{1}{4}} - 10x^{\frac{1}{4}} + x^{\frac{3}{4}} + 2x^{\frac{3}{4}} = (14 - 10)\,x^{\frac{1}{4}} + (1 + 2)\,x^{\frac{3}{4}} = 4x^{\frac{1}{4}} + 3x^{\frac{3}{4}}$$

Therefore, $14x^{\frac{1}{4}} + x^{\frac{3}{4}} - 10x^{\frac{1}{4}} + 2x^{\frac{3}{4}} = 4x^{\frac{1}{4}} + 3x^{\frac{3}{4}}$.

Simplify each of the following expressions.

13. $9x^5 - 4x^3 + 7x^3 - 8x^5$

14. $-4y^7 - y^6 + 12y^7 - 6y^6$

15. $22a^{\frac{4}{5}} - 13a^{\frac{3}{5}} - 5a^{\frac{4}{5}} + a^{\frac{4}{5}}$

16. $3x^{\frac{1}{8}} + 11y^{\frac{1}{8}} - 3y^{\frac{1}{8}} + 12x^{\frac{1}{8}}$

17. $-b^{\frac{5}{7}} + b^{\frac{2}{7}} + 2b^{\frac{5}{7}} - 7b^{\frac{2}{7}}$

18. $6b^{\frac{9}{10}} - 16a^{\frac{9}{10}} - 9a^{\frac{9}{10}} - 3b^{\frac{9}{10}}$

5.5   Inverses of Exponential Functions

A function in the form $y = a^x$, where a is a constant base, is called an **exponential function**.
To find the **inverse** of any function, the positions of x and y should be switched, so $y = a^x$ would
become $x = a^y$. Another way of writing $x = a^y$ is $y = \log_a x$, so $y = \log_a x$ is the inverse of
$y = a^x$. The graphs of the two functions are symmetrical with respect to the line $y = x$.
A function in the form $y = \log_a x$ is called a **logarithmic function**.
The logarithm with base e is called the **natural logarithm**, denoted by $\ln x$. $\log_e x = \ln x$.

Example 12:   What is the inverse of the function $y = 0.5^x$?

   Step 1:   Switch the positions of x and y.

   The function $y = 0.5^x$ is an exponential function with a base of 0.5. To find the
   inverse of the function, the first step is to switch the positions of x and y, so
   $y = 0.5^x$ would become $x = 0.5^y$.

   Step 2:   Rewrite the function.

   Another way of writing $x = 0.5^y$ is $y = \log_{0.5} x$, so $y = \log_{0.5} x$ is the inverse
   of $y = 0.5^x$.

Find the inverse of each of the following functions.

1. $y = 8^x$

2. $y = 0.25^x$

3. $y = e^x$

4. $y = 10^x$

5. $y = \left(\frac{1}{6}\right)^x$

6. $y = 175^x$

Example 13:   What is the inverse of the function $y = \log_7 x$?

   Step 1:   Rewrite the function.

   Another way of writing the function $y = \log_7 x$ is $x = 7^y$.

   Step 2:   Switch the positions of x and y.

   To find the inverse of any function, the first step is to switch the positions of x
   and y, so when finding the inverse of $x = 7^y$, it would first become $y = 7^x$. In
   this case, there are no additional operations to perform, so the inverse of $x = 7^y$
   is $y = 7^x$. This means that the inverse of $y = \log_7 x$ is also $y = 7^x$.

Find the inverse of each of the following functions.

7. $y = \log_{50} x$

8. $y = \log_{\frac{2}{9}} x$

9. $y = \log_{5.5} x$

10. $y = \log_{12} x$

11. $y = \log_{\frac{1}{10}} x$

12. $y = \log_{3.14} x$

Example 14:   The graphs of the functions $y = 2^x$, $y = x$, and $y = \log_2 x$ are shown below.

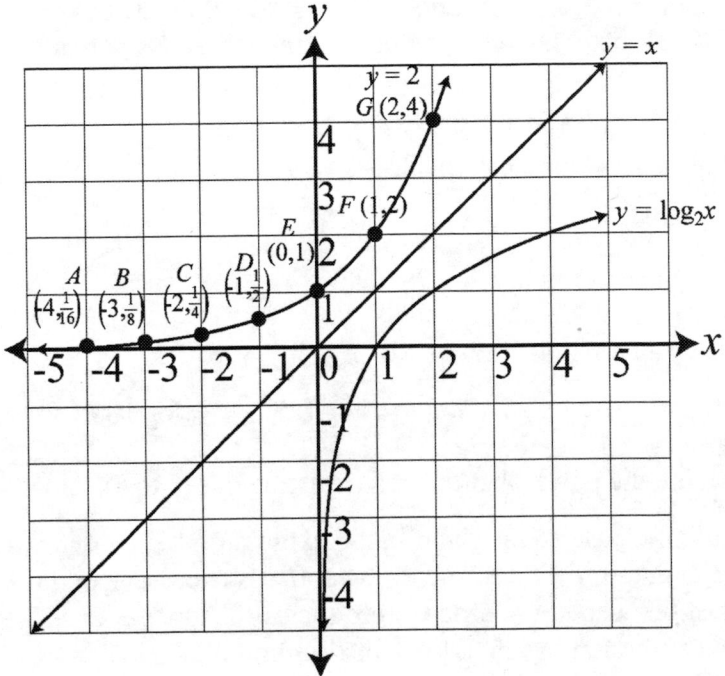

Find the coordinates of a point on the graph of that is a reflection across the line $y = x$ of point E on the graph of $y = 2^x$.

Step 1:   Find the coordinates of the point on the graph of $y = 2^x$.

The coordinates of point E on the graph of $y = 2^x$ are $(0, 1)$.

Step 2:   Reverse the coordinates of the point.

The graphs of the functions $y = 2^x$ and $y = \log_2 x$ are symmetrical with respect to the line $y = x$. This means that to find the coordinates of a point on the graph of $y = \log_2 x$ that is a a reflection across the line $y = x$ of point E on the graph of $y = 2^x$, all that needs to be done is to reverse the coordinates of point E. Since the coordinates of point E are $(0, 1)$, the coordinates of a point that is a reflection of point E are $(1, 0)$.

Find the coordinates of a point on the graph of $y = \log_2 x$ that is a reflection across the line $y = x$ of each of the following points. Use the graph above.

13. C 15. G 17. F

14. A 16. B 18. D

Copyright © American Book Company   81

5.6   Properties of Logarithms

If $a^m = c$ and $a^n = d$, then $\log_a c = m$ and $\log_a d = n$. A logarithm is just an exponent, so logarithmic properties can all be obtained by extending the laws of exponents.

Basic Properties of Logarithms

$$\log_a (cd) = \log_a c + \log_a d \qquad \log_a \left(\frac{c}{d}\right) = \log_a c - \log_a d \qquad \log_a \left(c^d\right) = d \log_a c$$

Example 15:   Simplify the expression $\log 25 + \log 4$.

Step 1:   Use the property $\log_a (cd) = \log_a c + \log_a d$ or $\log_a c + \log_a d = \log_a (cd)$.

This means that $\log 25 + \log 4 = \log (25 \cdot 4) = \log 100$.

Step 2:   Determine the power that the base must be raised to.

Because the base is not given in the expression $\log 100$, it is understood to be 10, so to determine the value of $\log 100$, it's necessary to think of the power that 10 must be raised to in order to produce 100. The power is 2, since $10^2 = 100$, so the expression $\log 25 + \log 4$ in simplified form is 2.

Simplify each of the following expressions.

1. $\log_4 8 + \log_4 8$

2. $\log_2 0.1 + \log_2 160$

3. $\log 2 + \log 5$

4. $\log_6 24 + \log_6 9$

5. $\log_9 2 + \log_9 4.5$

6. $\log_{15} 75 + \log_{15} 3$

Example 16:   Simplify the expression $\log_5 1000 - \log_5 8$.

Step 1:   Use the property $\log_a \left(\frac{c}{d}\right) = \log_a c - \log_a d$ or $\log_a c - \log_a d = \log_a \left(\frac{c}{d}\right)$.

This means that $\log_5 1000 - \log_5 8 = \log_5 \left(\frac{1000}{8}\right) = \log_5 125$.

Step 2:   Determine the power that the base must be raised to.

To determine the value of $\log_5 125$, it's necessary to think of the power that 5 must be raised to in order to produce 125. The power is 3, since $5^3 = 125$, so the expression $\log_5 1000 - \log_5 8$ in simplified form is 3.

Simplify each of the following expressions.

7. $\log_3 486 - \log_3 2$

8. $\log_7 294 - \log_7 6$

9. $\log_8 88 - \log_8 11$

10. $\log 9 - \log 90$

11. $\log_2 448 - \log_2 7$

12. $\log_5 3 - \log_5 75$

Example 17:   Simplify the expression $-8\log_{16} 2$.

Step 1:   Use the property $\log_a\left(c^d\right) = d\log_a c$.

Since $\log_a\left(c^d\right) = d\log_a c$, it is also true that $d\log_a c = \log_a\left(c^d\right)$. This means that $-8\log_{16} 2 = \log_{16}\left(2^{-8}\right) = \log_{16}\left(\frac{1}{256}\right)$.

Step 2:   Determine the power that the base must be raised to.

To determine the value of $\log_{16}\left(\frac{1}{256}\right)$, it's necessary to think of the power that 16 must be raised to in order to produce $\frac{1}{256}$. The power is -2, since $16^{-2} = \frac{1}{256}$, so the expression $-8\log_{16} 2$ in simplified form is -2.

Simplify each of the following expressions.

13. $6\log_9 3$

14. $2\log_{25}\left(\frac{1}{5}\right)$

15. $125\log 1$

16. $-3\log_8 4$

17. $10\log_4 2$

18. $-5\log_{32}\left(\frac{1}{2}\right)$

Example 18:   Write the expression $4\log_7 6 - \log_7 8$ as a single logarithm.

Step 1:   Use the property $\log_a\left(c^d\right) = d\log_a c$.

To write the expression $4\log_7 6 - \log_7 8$ as a single logarithm $4\log_7 6$ should be rewritten as $\log_7 6^4$, which is equal to $\log_7 1296$. The expression then becomes $\log_7 1296 - \log_7 8$.

Step 2:   Use the property $\log_a\left(cd\right) = \log_a c + \log_a d$ or the property $\log_a\left(\frac{c}{d}\right) = \log_a c - \log_a d$.

Next, $\log_7 1296 - \log_7 8$ rewritten as $\log_7\left(\frac{1296}{8}\right)$, which is equal to $\log_7 162$. This means that the expression $4\log_7 6 - \log_7 8$ written as a single logarithm is $\log_7 162$.

Write each of the following expressions as a single logarithm.

19. $2\log_5 14 - \log_5 2$

20. $\log_{11} 9 + 3\log_{11} 4$

5.7   Characteristics of Exponential and Logarithmic Functions

A function in the form $y = ax$, where a is a constant base, is called an exponential function, and one in the form $y = \log_a x$ is called a logarithmic function. Exponential and logarithmic functions and their graphs have certain characteristics that include domain and range, zeros, asymptotes, x and y-intercepts, intervals of increase and decrease, and rate of change. The intervals of increase and decrease and the rate of change for both types of functions depend on whether the base a is greater than 1 or between 0 and 1. The characteristics of both types of functions are summarized in the table below.

	Exponential Functions $(y = a^x)$	**Logarithmic Functions** $(y = \log_a x)$
Domain	All real numbers	All real numbers greater than 0
Range	All real numbers greater than 0	All real numbers
Zeros	None	$x = 1$
Asymptotes	The x-axis	The y-axis
x-intercepts	None	$(1, 0)$
y-intercepts	$(0, 1)$	None
Intervals of Increase and Decrease	If $a > 1$, the function increases throughout its domain. If $0 < a < 1$, it decreases throughout its domain.	If $a > 1$, the function increases throughout its domain. If $0 < a < 1$, it decreases throughout its domain.
Rate of Change	If $a > 1$, the rate of change is positive and moves away from 0. If $0 < a < 1$, it's negative and approaches 0.	If $a > 1$, the rate of change is positive and approaches 0. If $0 < a < 1$, it's negative and approaches 0.

Example 19:   The graph below is that of either an exponential or a logarithmic function. Give the domain, range, and zeros of the function, as well as the graph's asymptotes, x-intercepts, y-intercepts, intervals of increase and decrease, and rate of change. Then state whether the graph is that of an exponential function or a logarithmic function and whether the base is greater than 1 or between 0 and 1.

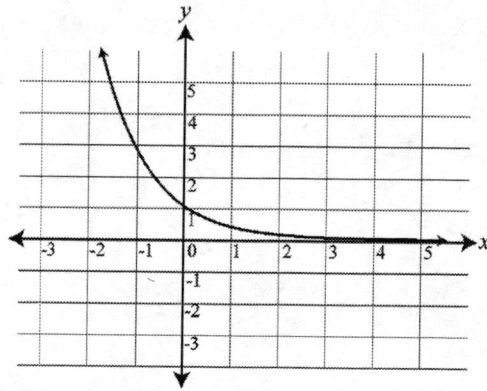

Copyright © American Book Company

Step 1: Determine the **domain** and **range** of the function.

Because the function is either an exponential function or a logarithmic function, either its domain is all real numbers and its range is all real numbers greater than 0, or vice versa. Since the graph is on both sides of the y-axis but only on the positive side of the x-axis, the function's domain is all real numbers and its range is all real numbers greater than 0.

Step 2: Determine the **zeros** of the function.

Also, an exponential function has no zeros, while a logarithmic function has a zero of $x = 1$. Since the graph does not pass through the x-axis at $x = 1$, the function must have no zeros.

Step 3: Determine the graph's **asymptotes**.

The graph of the function is getting closer and closer to the x-axis, but it never touches the x-axis. This means that the x-axis is an asymptote of the graph.

Step 4: Determine the x-**intercepts** and y-**intercepts** of the graph.

Also, the graph passes through the y-axis at the point $(0, 1)$, but it never passes through the x-axis. This means that the graph doesn't have an x-intercept, but it has a y-intercept of $(0, 1)$.

Step 5: Determine the graph's **intervals of increase and decrease** and **rate of change**.

In addition, moving from left to right, the graph is going down and leveling off. This means that the function is decreasing throughout its domain and that its rate of change is negative and approaching 0.

Step 6: State whether the graph is that of an exponential function or a logarithmic function and whether the base is greater than 1 or between 0 and 1.

Therefore, the graph must be that of an exponential function, and the base must be between 0 and 1.

Use the graph below to determine each of the following.

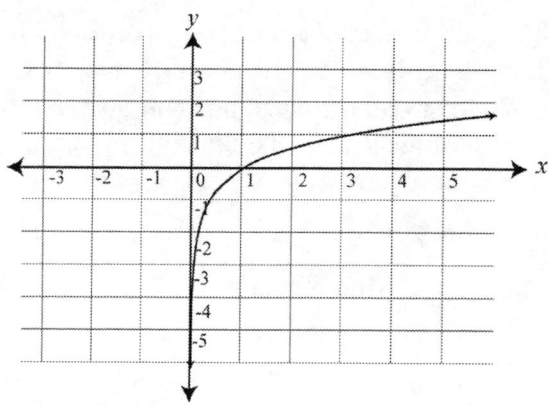

1. Find the domain of the function.

2. Find the range of the function.

3. Find the zeros of the function.

4. Find the graph's asymptotes.

5. Find the graph's x-intercepts.

6. Find the graph's y-intercepts.

7. Find the graph's intervals of increase and decrease.

8. Find the graph's rate of change.

9. Is the function exponential or logarithmic?

10. Is the base greater than 1 or between 0 and 1?

Example 20:   The graph below is that of either an exponential or a logarithmic function. Give the domain, range, and zeros of the function, as well as the graph's asymptotes, x-intercepts, y-intercepts, intervals of increase and decrease, and rate of change. Then state whether the graph is that of an exponential function or a logarithmic function and whether the base is greater than 1 or between 0 and 1.

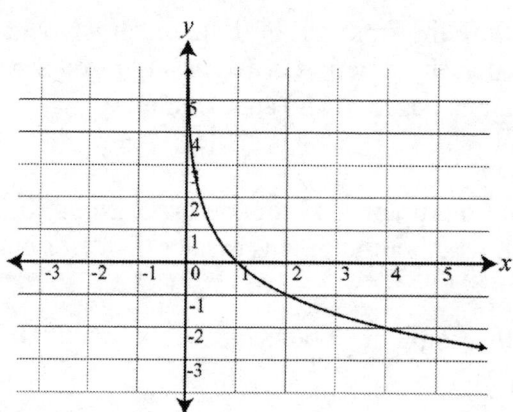

Step 1:   Determine the **domain** and **range** of the function.
Because the function is either an exponential function or a logarithmic function, either its domain is all real numbers and its range is all real numbers greater than 0, or vice versa. Since the graph is on both sides of the x-axis but only on the positive side of the y-axis, the function's domain is all real numbers greater than 0 and its range is all real numbers.

Step 2:   Determine the **zeros** of the function.
Also, an exponential function has no zeros, while a logarithmic function has a zero of $x = 1$. Since the graph passes through the x-axis at $x = 1$, the function must have a zero of $x = 1$.

Step 3:   Determine the graph's **asymptotes**.
The graph of the function is getting closer and closer to the y-axis, but it never touches the y-axis. This means that the y-axis is an asymptote of the graph.

Step 4:   Determine the x-**intercepts** and y-**intercepts** of the graph.
Also, even though the graph passes through the x-axis at the point $(1, 0)$, it never passes through the y-axis. This means that the graph doesn't have a y-intercept, but it has an x-intercept of $(1, 0)$.

Step 5:   Determine the graph's **intervals of increase and decrease** and **rate of change**.
In addition, moving from left to right, the graph is going down and leveling off. This means that the function is decreasing throughout its domain and that its rate of change is negative and approaching 0.

Step 6:   State whether the graph is that of an exponential function or a logarithmic function and whether the base is greater than 1 or between 0 and 1.
Therefore, the graph must be that of a logarithmic function, and the base must be between 0 and 1.

Use the graph below to determine each of the following.

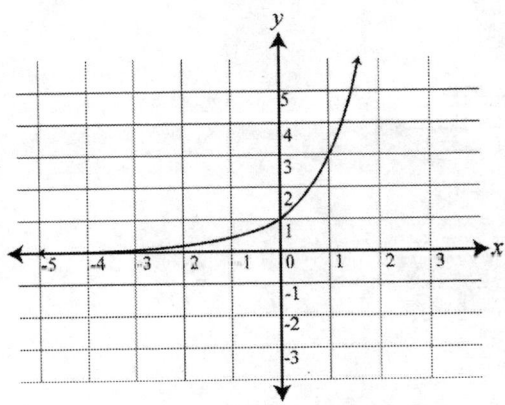

11.   Find the domain of the function.
12.   Find the range of the function.
13.   Find the zeros of the function.
14.   Find the graph's asymptotes.
15.   Find the graph's x-intercepts.
16.   Find the graph's y-intercepts.
17.   Find the graph's intervals of increase and decrease.
18.   Find the graph's rate of change.
19.   Is the function exponential or logarithmic?
20.   Is the base greater than 1 or between 0 and 1?

5.8   Transformations of Logarithmic and Exponential Functions

The graphs of the functions $f(x) = a^x$, $f(x) = \log_a x$, $f(x) = e^x$, and $f(x) = \ln x$ can be transformed. Each of the transformations is described in the table below.

Transformation	Transformed Functions	Conditions
Vertical Stretch or Compression	$f(x) = c \cdot a^x$, $f(x) = c \cdot \log_a x$, $f(x) = c \cdot e^x$, $f(x) = c \cdot \ln x$	c is a constant that has an absolute value greater than 1 for a vertical stretch and an absolute value less than 1 for a vertical compression
Reflection Across the x-axis	$f(x) = c \cdot a^x$, $f(x) = c \cdot \log_a x$, $f(x) = c \cdot e^x$, $f(x) = c \cdot \ln x$	c is a constant that is negative
Horizontal Stretch or Compression	$f(x) = a^{d \cdot x}$, $f(x) = \log_a(d \cdot x)$, $f(x) = e^{d \cdot x}$, $f(x) = \ln(d \cdot x)$	d is a constant that has an absolute value less than 1 for a horizontal stretch and an absolute value greater than 1 for a horizontal compression
Reflection Across the y-axis	$f(x) = a^{x-h}$, $f(x) = \log_a(d \cdot x)$, $f(x) = e^{d \cdot x}$, $f(x) = \ln(d \cdot x)$	d is a constant that is negative
Translation Right or Left	$f(x) = a^{x-h}$, $f(x) = \log_a(x - h)$, $f(x) = e^{x-h}$, $f(x) = \ln(x - h)$	h is a constant that is positive for a translation right and negative for a translation left
Translation Up or Down	$f(x) = a^x + k$, $f(x) = \log_a x + k$, $f(x) = e^x + k$, $f(x) = \ln x + k$	k is a constant that is positive for a translation up and negative for a translation down

Example 21:   The graph of the function $f(x) = e^x$ was transformed by a vertical stretch, a vertical compression, a reflection across the x-axis, or a combination of these transformations to produce the graph of the function $f(x) = -0.25e^x$ as shown below. Determine the transformations that were applied.

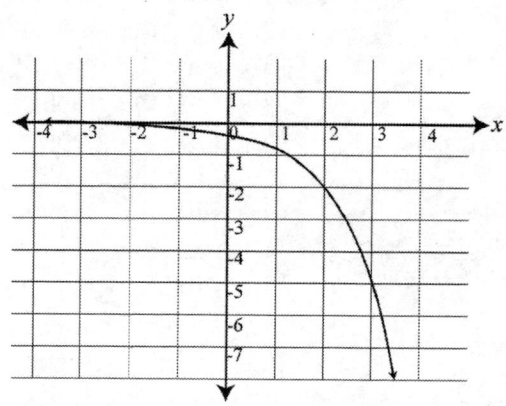

Step 1: Check for a vertical stretch or compression.

The function $f(x) = -0.25e^x$ is in the form $f(x) = c \cdot e^x$, with the value of c being -0.25. Since the absolute value of -0.25 is less than 1, the graph of the function $f(x) = e^x$ has undergone a vertical compression.

Step 2: Check for a reflection across the x-axis.

Also, since c is negative, the graph of the function $f(x) = e^x$ has undergone a reflection across the x-axis.

The function $f(x) = \log_4 x$ was transformed by a vertical stretch, a vertical compression, a reflection across the x-axis, or a combination of these to produce the graphs of each of the following functions. Determine the transformations that were applied to each.

1. $f(x) = 8 \log_4 x$ 3. $f(x) = -\frac{3}{4} \log_4 x$ 5. $f(x) = -22 \log_4 x$

2. $f(x) = -3 \log_4 x$ 4. $f(x) = 0.9 \log_4 x$ 6. $f(x) = \frac{7}{10} \log_4 x$

Example 22: The graph of the function $f(x) = \log x$ was transformed by a horizontal stretch, a horizontal compression, a reflection across the y-axis, or a combination of these transformations to produce the graph of the function $f(x) = \log(6x)$ as shown below. Determine the transformations that were applied.

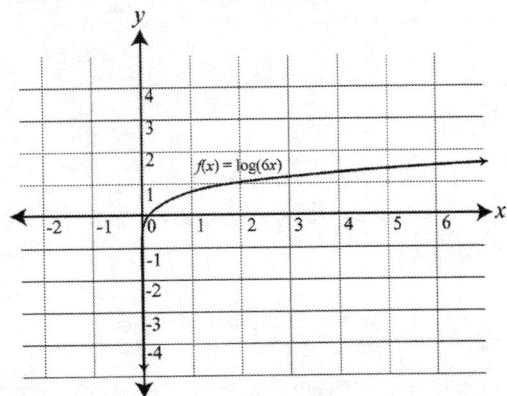

Step 1: Check for a horizontal stretch or compression.

The function $f(x) = \log(6x)$ is in the form $\log_a(d \cdot x)$, with the value of d being 6. Since the absolute value of 6 is greater than 1, the graph of the function $f(x) = \log x$ has undergone a horizontal compression.

Step 2: Check for a reflection across the y-axis.

Also, since d is positive, the graph of the function $f(x) = \log x$ has not undergone a reflection across the y-axis.

The function $f(x) = 14^x$ was transformed by a horizontal stretch, a horizontal compression, a reflection across the y-axis, or a combination of these to produce the graphs of each of the following functions. Determine the transformations that were applied to each.

7. $f(x) = 14^{-5x}$

9. $f(x) = 14^{14x}$

11. $f(x) = 14^{\frac{4x}{5}}$

8. $f(x) = 14^{0.1x}$

10. $f(x) = 14^{\frac{x}{3}}$

12. $f(x) = 14^{-x}$

Example 23:   The graph of the function $f(x) = \ln x$ was transformed by a translation right or left, a translation up or down, or a combination of these transformations to produce the graph of the function $f(x) = \ln(x + 3) - 2$ as shown below. Determine the transformations that were applied.

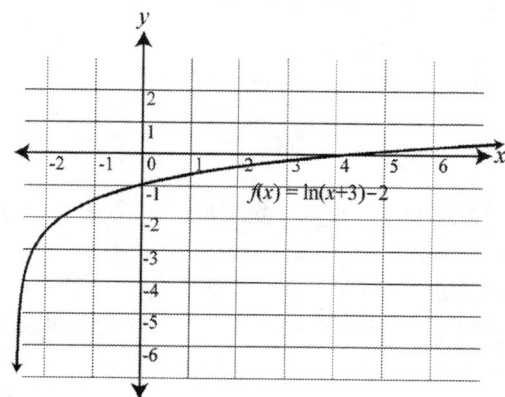

$f(x) = \ln(x+3) - 2$

Step 1:   Check for a translation right or left.

The function $f(x) = \ln(x + 3) - 2$ is in the form $f(x) = \ln(x - h) + k$, with the value of h being -3. Since h is negative, the graph of the function $f(x) = \ln x$ has undergone a translation left.

Step 2:   Check for a translation up or down.

Also, the value of k is -2. Since k is negative, the graph of the function $f(x) = \ln x$ has undergone a translation down.

The function $f(x) = 0.8^x$ was transformed by a translation right or left, a translation up or down, or a combination of these to produce the graphs of each of the following functions. Determine the transformations that were applied to each.

13. $f(x) = 0.8^x + 11$

15. $f(x) = 0.8^{x+0.8}$

17. $f(x) = 0.8^x - 8$

14. $f(x) = 0.8^{x-5} - 9$

16. $f(x) = 0.8^{x+6} + 4$

18. $f(x) = 0.8^{x-20}$

5.9   Real-World Exponential and Logarithmic Functions

Many problems in the real world can be solved with exponential and logarithmic functions. These problems involve exponential decay and exponential growth.

Exponential decay occurs when a quantity decreases at a rate proportional to its value, while **exponential growth** occurs when a quantity increases at a rate proportional to its value.

When there is exponential decay, it's possible to calculate a quantity's half-life, or the time it takes for the quantity to be cut in half, and when there is exponential growth, it's possible to calculate a quantity's doubling time, or the time it takes for the quantity to double. Also, given a quantity's half-life or doubling time, the quantity's value at some point in the future can be calculated if its starting value is known.

Example 24:   The half-life of aspirin's presence in the human body is 3.2 hours. If a person takes 650 mg of aspirin, about how many milligrams will be present in the person's body 16 hours after taking it? Round your answer to the nearest milligram.

Step 1:   Substitute values into the formula $Q = Q_0 \left(\frac{1}{2}\right)^{\frac{t}{h}}$, where Q is the future value of the quantity, Q_0 is the quantity's starting value, h is the half-life of the quantity, and t is the time elapsed.

When the formula $Q = Q_0 \left(\frac{1}{2}\right)^{\frac{t}{h}}$ is used to calculate the number of milligrams of aspirin present, $Q_0 = 650$ mg, $h = 3.2$ hours, and $t = 16$ hours. The values of the variables can be substituted into the formula as follows.

$$Q = Q_0 \left(\frac{1}{2}\right)^{\frac{t}{h}} = 650 \left(\frac{1}{2}\right)^{\frac{16}{3.2}}$$

Step 2:   Solve the equation for Q.

$$Q = 650 \left(\frac{1}{2}\right)^{\frac{16}{3.2}} = 650 \left(\frac{1}{2}\right)^{5} = 650 \left(\frac{1}{32}\right) = \frac{650}{32} = 20.3125$$

This means that there will be about 20 mg of aspirin in the person's body 16 hours after taking it.

The starting value, half life, and time elapsed of various substances' presence in the human body are given. Calculate the approximate amount of each substance present after the given time elapsed. Round your answers to the nearest milligram.

1. Starting value = 400 mg; Half-life = 15 minutes; Time elapsed = 45 minutes

2. Starting value = 250 mg; Half-life = 4.5 hours; Time elapsed = 18 hours

3. Starting value = 575 mg; Half-life = 2 days; Time elapsed = 12 days

4. Starting value = 725 mg; Half-life = 7 minutes; Time elapsed = 56 minutes

5. Starting value = 150 mg; Half-life = 3 hours; Time elapsed = 21 hours

6. Starting value = 600 mg; Half-life = 4 days; Time elapsed = 8 days

Example 25:   Meg deposited $2,500 into a savings account offering continuously compounded interest. At the current interest rate, it will take 24 years for her money to double. If the interest rate stays the same and Meg doesn't make any additional deposits or withdrawals, how much money will she have in her savings account after 6 years?

Step 1:   Substitute values into the formula $Q = Q_0 \left(2\right)^{\frac{t}{d}}$, where Q is the future value of the quantity, Q_0 is the quantity's starting value, d is the doubling time of the quantity, and t is the time elapsed.

When the formula $Q = Q_0 \left(2\right)^{\frac{t}{d}}$ is used to calculate the amount of money in Meg's savings account, $Q_0 = \$2,500$, $d = 24$ years, and $t = 6$ years. The values of the variables can be substituted into the formula as follows.

$$Q = Q_0 \left(2\right)^{\frac{t}{d}} = 2,500 \left(2\right)^{\frac{6}{24}}$$

Step 2:   Solve the equation for Q.

$$Q = 2,500 \left(2\right)^{\frac{6}{24}} = 2,500 \left(2\right)^{\frac{1}{4}} = 2,500 \left(1.189207\right) = 2,973.02$$

This means that Meg will have $2,973.02 in her savings account after 6 years.

The starting value, doubling time, and time elapsed of various savings account deposits are given. Calculate the amount of money in each savings account after the given time elapsed. Assume that no additional deposits or withdrawals are made and that interest is compounded continuously.

7. Starting value = $6,000; Doubling time = 18 years; Time elapsed = 3 years

8. Starting value = $3,750; Doubling time = 12 years; Time elapsed = 4 years

9. Starting value = $7,500; Doubling time = 8 years; Time elapsed = 24 years

10. Starting value = $2,250; Doubling time = 36 years; Time elapsed = 9 years

11. Starting value = $4,500; Doubling time = 9 years; Time elapsed = 18 years

12. Starting value = $9,000; Doubling time = 6 years; Time elapsed = 30 years

Example 26:   It takes 30 years for 500 grams of a certain radioactive isotope to decay to 350 grams. What is the isotope's half life? Round your answer to the nearest year.

Step 1:   Substitute values into the formula $Q = Q_0 \left(\frac{1}{2}\right)^{\frac{t}{h}}$, where Q is the future value of the quantity, Q_0 is the quantity's starting value, h is the half-life of the quantity, and t is the time elapsed.

$$Q = Q_0 \left(\frac{1}{2}\right)^{\frac{t}{h}}, Q_0 = 500, t = 30, Q = 350$$

$$350 = 500 \left(\frac{1}{2}\right)^{\frac{30}{h}}$$

Step 2:   Divide each side of the equation by the quantity's starting value.

$$\frac{350}{500} = \left(\frac{1}{2}\right)^{\frac{30}{h}}$$

Step 3:   Take the natural logarithm of both sides of the equation and use the property $\log_a \left(c^d\right) = d \log_a c$.

$$\ln \left(\frac{350}{500}\right) = \ln \left(\left(\frac{1}{2}\right)^{\frac{30}{h}}\right) = \frac{30}{h} \times \ln \left(\frac{1}{2}\right)$$

Step 4:   Solve the equation $\ln \left(\frac{350}{500}\right) = \frac{30}{h} \times \ln \left(\frac{1}{2}\right)$ for h.

$$\ln \left(\frac{350}{500}\right) h = 30 \times \ln \left(\frac{1}{2}\right)$$

$$h = \frac{30 \times \ln \left(\frac{1}{2}\right)}{\ln \left(\frac{350}{500}\right)} = 58.3007$$

This means that the isotope's half-life is about 58 years.

The future value, starting value, and time elapsed of various radioactive isotopes are given. Calculate the half life of each of the isotopes. Round your answers to the nearest year.

13. Future value = 200 grams; Starting value = 275 grams; Time elapsed = 10 years

14. Future value = 675 grams; Starting value = 800 grams; Time elapsed = 25 years

15. Future value = 450 grams; Starting value = 725 grams; Time elapsed = 55 years

Example 27:   The population of a country is expected to increase exponentially from $40,000,000$ people to $64,000,000$ people in a span of 9 years. At this rate of increase, what is the approximate doubling time of the country's population? Round your answer to the nearest year.

Step 1:   Substitute values into the formula $Q = Q_0 \left(2\right)^{\frac{t}{d}}$, where Q is the future value of the quantity, Q_0 is the quantity's starting value, d is the doubling time of the quantity, and t is the time elapsed.

$Q = 64,000,000$ people, $Q_0 = 40,000,000$ people, and $t = 9$ years. Substitute the known values into the formula.

$$Q = Q_0 \left(2\right)^{\frac{t}{d}}$$

$$64,000,000 = 40,000,000 \left(2\right)^{\frac{9}{d}}$$

Step 2:   Divide each side of the equation by the quantity's starting value.

$$\frac{64,000,000}{40,000,000} = \left(2\right)^{\frac{9}{d}}$$

Step 3:   Take the natural logarithm of both sides of the equation and use the property $\log_a\left(c^d\right) = d\log_a c$.

$$\ln\left(\frac{64,000,000}{40,000,000}\right) = \ln\left(\left(2\right)^{\frac{9}{d}}\right)$$

$$\ln\left(\frac{64,000,000}{40,000,000}\right) = \frac{9}{d} \times \ln\left(2\right)$$

Step 4:   Solve the equation for d.

$$\ln\left(\frac{64,000,000}{40,000,000}\right) d = 9 \times \ln\left(2\right)$$

$$d = \frac{9 \times \ln\left(2\right)}{\ln\left(\frac{64,000,000}{40,000,000}\right)} = 13.2729$$

This means that the doubling time of the country's population is about 13 years.

The future value, starting value, and time elapsed of the populations of various countries are given. Calculate the doubling time of each of the populations. Round your answers to the nearest year.

16. Future value = 91 million; Starting value = 70 million; Time elapsed = 7 years

17. Future value = 55 million; Starting value = 50 million; Time elapsed = 4 years

18. Future value = 36 million; Starting value = 24 million; Time elapsed = 19 years

Chapter 5 Review

Find each of the following roots.

1. $\sqrt[4]{16}$

2. $\sqrt[3]{-729}$

3. $\sqrt[5]{243}$

Solve each of the following equations for x.

4. $\sqrt[6]{8} \times \sqrt[6]{8} = x$

5. $\dfrac{\sqrt[3]{1024}}{\sqrt[3]{2}} = x$

6. $\sqrt[7]{3^{13}} = 3^x$

Solve each of the following equations for x.

7. $7^{\frac{3}{10}} \times 7^{\frac{1}{5}} = 7^x$

8. $\dfrac{12^{\frac{5}{6}}}{12^{\frac{2}{3}}} = 12^x$

9. $\left(15^3\right)^4 = 15^x$

Simplify each of the following expressions.

10. $2x^8 - 5x^9 + 11x^8 - 6x^9$

11. $7x^{\frac{8}{9}} + 14x^{\frac{1}{9}} - 4x^{\frac{8}{9}} + x^{\frac{1}{9}}$

12. $10x^{\frac{5}{8}} - 8y^{\frac{5}{8}} + 2x^{\frac{5}{8}} - 17y^{\frac{5}{8}}$

Find the inverse of each of the following functions.

13. $y = 14^x$

14. $y = \left(\frac{3}{4}\right)^x$

15. $y = 0.9^x$

16. $y = \log_7 x$

17. $y = \log_{\frac{1}{3}} x$

18. $y = \log_{2.5} x$

Simplify each of the following expressions.

19. $\log_5 12.5 + \log_5 2$

20. $\log_2 176 - \log_2 11$

21. $\log_{12} 3 + \log_{12} 4$

22. $4 \log_{16} 4$

23. $\log_6 864 - \log_6 4$

24. $-2 \log_4 8$

Use the graph below to determine each of the following.

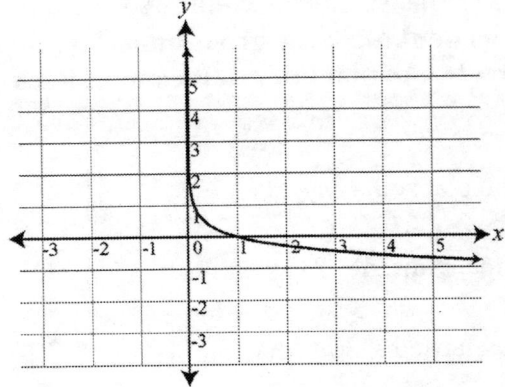

25. Find the domain and range of the function.
26. Find the zeros of the function.
27. Find the graph's asymptotes.
28. Find the x-intercepts and y-intercepts of the graph.
29. Find the graph's intervals of increase and decrease and rate of change.
30. Is the graph that of an exponential function or a logarithmic function and is the base greater than 1 or between 0 and 1?

The function $f(x) = 5^x$ was transformed by a horizontal or vertical stretch, a horizontal or vertical compression, a reflection across the x-axis or y-axis, or a combination of these to produce the graphs of each of the following functions. Determine the transformations that were applied to each.

31. $f(x) = -2(5^x)$

32. $f(x) = 5^{\frac{x}{9}}$

33. $f(x) = \frac{1}{4}(5^{6x})$

The function $f(x) = \log_{20} x$ was transformed by a translation right or left, a translation up or down, or a combination of these to produce the graphs of each of the following functions. Determine the transformations that were applied to each.

34. $f(x) = \log_{20}(x - 1) - 18$

35. $f(x) = \log_{20}(x + 4) - 4$

36. $f(x) = \log_{20} x + 20$

The future value, starting value, and time elapsed of various radioactive isotopes are given. Calculate the half-life of each of the isotopes. Round your answers to the nearest year.

37. Future value = 700 grams; Starting value = 950 grams; Time elapsed = 28 years

38. Future value = 150 grams; Starting value = 175 grams; Time elapsed = 13 years

39. Future value = 300 grams; Starting value = 425 grams; Time elapsed = 44 years

The starting value, doubling time, and time elapsed of various savings account deposits are given. Calculate the amount of money in each savings account after the given time elapsed. Assume that no additional deposits or withdrawals are made and that interest is compounded continuously.

40. Starting value = $9,500; Doubling time = 15 years; Time elapsed = 5 years

41. Starting value = $12,750; Doubling time = 27 years; Time elapsed = 21 years

42. Starting value = $8,000; Doubling time = 9 years; Time elapsed = 30 years

Chapter 5 Test

1. What is $\sqrt[3]{-64}$?

   A. -8
   B. -4
   C. 4
   D. 8

2. Which of these expressions is equal to 6?

   A. $\dfrac{\sqrt[4]{108}}{\sqrt[4]{12}}$

   B. $\dfrac{\sqrt{108}}{\sqrt{12}}$

   C. $\sqrt[4]{108} \times \sqrt[4]{12}$

   D. $\sqrt{108} \times \sqrt{12}$

3. Which of these statements is true?

   A. The nth root of an odd number is always negative.
   B. The nth root of an odd number is always positive.
   C. The nth root of an odd number is always even.
   D. The nth root of an odd number is always odd.

4. Which of these is a valid equation for all values of a and b?

   A. $\sqrt{a+b} = \sqrt{a} + \sqrt{b}$
   B. $\sqrt{a+b} = \sqrt{a} \times \sqrt{b}$
   C. $\sqrt{ab} = \sqrt{a} + \sqrt{b}$
   D. $\sqrt{ab} = \sqrt{a} \times \sqrt{b}$

5. What is $8^{\frac{1}{4}} \times 8^{\frac{3}{20}}$?

   A. $8^{\frac{1}{5}}$

   B. $8^{\frac{2}{5}}$

   C. $8^{\frac{3}{5}}$

   D. $8^{\frac{4}{5}}$

6. Which of these expressions is equal to a^8?

   A. $a^4 \div a^2$
   B. $a^6 \div a^2$
   C. $a^8 \div a^2$
   D. $a^{10} \div a^2$

7. If $\dfrac{15^9}{15^3} = 15^x$, what is the value of x?

   A. 3
   B. 6
   C. 12
   D. 27

8. The expression $\left(b^4\right)^4$ is the same as which of these expressions?

   A. $b^4 + b^4$
   B. $b^4 + b^4 + b^4 + b^4$
   C. $b^4 \times b^4$
   D. $b^4 \times b^4 \times b^4 \times b^4$

9. What is another way of writing $x = a^y$?

   A. $x = \log_a y$
   B. $x = \log_a a$
   C. $y = \log_a x$
   D. $y = \log_x a$

10. What is the inverse of the function $y = \log_{11} x$?

    A. $x = y^{11}$
    B. $x = 11^y$
    C. $y = x^{11}$
    D. $y = 11^x$

11. What is the base of the function $y = \ln x$?

    A. 1
    B. The constant e
    C. The constant π
    D. 10

12. The graphs of function and its inverse are symmetrical with respect to which of the following lines?

   A. The x-axis

   B. The y-axis

   C. The line $y = -x$

   D. The line $y = x$

13. Which of these expressions is equal to 2?

   A. $\log 50 - \log 2$

   B. $\log 50 + \log 2$

   C. $\log 50 \times \log 2$

   D. $\log 50 \div \log 2$

14. The expression $\ln c - \ln d$ is equal to which of the following?

   A. $\ln (c - d)$

   B. $\ln (c + d)$

   C. $\ln (c \times d)$

   D. $\ln (c \div d)$

15. What is $-4 \log_9 3$?

   A. -8

   B. -2

   C. 2

   D. 8

16. Which of these $3 \log_6 8 - \log_6 32$ written as a single logarithm?

   A. $\log_6 \left(\frac{3}{8} \right)$

   B. $\log_6 \left(\frac{3}{4} \right)$

   C. $\log_6 8$

   D. $\log_6 16$

17. What is the domain and range of the exponential function $y = 7^x$?

   A. The domain is all real numbers, and the range is all real numbers.

   B. The domain is all real numbers, and the range is all real numbers greater than 0.

   C. The domain is all real numbers greater than 0, and the range is all real numbers.

   D. The domain is all real numbers greater than 0, and the range is all real numbers greater than 0.

18. Which of these is an asymptote of the graph of the function $y = \log_{12} x$?

   A. The x-axis

   B. The y-axis

   C. The line $y = -x$

   D. The line $y = x$

19. Which of these described the rate of change of the graph of the function $y = e^x$?

   A. It's negative and approaches 0.

   B. It's negative and moves away from 0.

   C. It's positive and approaches 0.

   D. It's positive and moves away from 0.

20. The graph below is that of either an exponential or a logarithmic function.

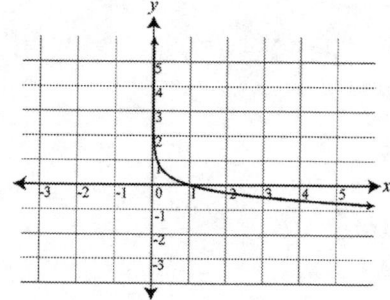

Which of these functions could it be?

   A. $y = \log_{\frac{1}{5}} x$

   B. $y = \log_5 x$

   C. $y = \left(\frac{1}{5} \right)^x$

   D. $y = 5^x$

21. If the function $f(x) = \log_7 x$ were transformed to produce the graph of the function $f(x) = -15 \log_7 x$, which of these transformations were applied?

   A. A vertical compression and a reflection across the x-axis

   B. A vertical compression and a reflection across the y-axis

   C. A vertical stretch and a reflection across the x-axis

   D. A vertical stretch and a reflection across the y-axis

22. A transformation was applied to the graph of the function $f(x) = e^x$ to produce the graph shown below.

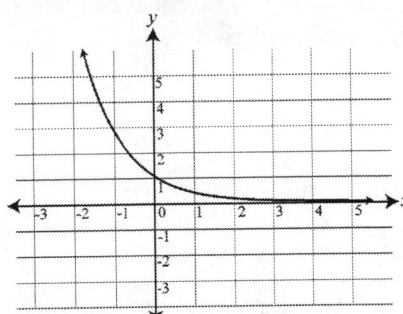

Which of these could be the equation of the graph?

   A. $f(x) = -e^x$

   B. $f(x) = e^{-x}$

   C. $f(x) = e^x - 1$

   D. $f(x) = e^x + 1$

23. If the graph of the function $f(x) = \ln x$ were translated 5 units to the right and 10 units down, what would be the equation of the resulting graph?

   A. $f(x) = \ln(x - 5) - 10$

   B. $f(x) = \ln(x - 5) + 10$

   C. $f(x) = \ln(x + 5) - 10$

   D. $f(x) = \ln(x + 5) + 10$

24. Transformations were applied to the graph of the function $f(x) = 60^x$ to produce each of the following functions. Which of them has undergone a horizontal compression?

   A. $f(x) = \left(\frac{1}{3}\right)(60^x)$

   B. $f(x) = 60^{\frac{x}{3}}$

   C. $f(x) = 60^{3x}$

   D. $f(x) = (3)(60^x)$

25. A medicine's half-life in the human body is 9 hours. If Tiana took 800 mg of the medicine, how much of it will be left in her body after 36 hours?

   A. 25 mg

   B. 50 mg

   C. 100 mg

   D. 200 mg

26. A country's population is growing exponentially and doubles every 10 years. If it is currently $10,000,000$ people, about what will it be in 5 years?

   A. $14,142,136$ people

   B. $15,000,000$ people

   C. $15,157,166$ people

   D. $20,000,000$ people

27. It takes 75 years for 900 grams of a certain radioactive isotope to decay to 700 grams. What is the isotope's half-life to the nearest year?

   A. 96 years

   B. 207 years

   C. 253 years

   D. 326 years

28. When does exponential decay occur?

   A. when a quantity decreases at a constant rate

   B. when a quantity increases at a constant rate

   C. when a quantity decreases at a rate proportional to its value

   D. when a quantity increases at a rate proportional to its value

Chapter 6
Solving Polynomial, Exponential, and Logarithmic Equations

This chapter covers the following Georgia Performance Standards:

MM3A	Algebra	MM3A3a, MM3A3b, MM3A3d

6.1 Factoring By Grouping

Polynomials of higher degree that have only one variable can be factored by grouping. First, they must be arranged in descending order. In the example below, you could try to factor it by grouping without rearranging it in descending order, but you will see it does not work.

Example 1: $2n^3 - 8 + 8n^2 - 2n$

Step 1: Arrange in descending order (exponents go from highest to lowest).
$2n^3 - 8 + 8n^2 - 2n = 2n^3 + 8n^2 - 2n - 8$

Step 2: Factor by grouping the first two terms and the last two terms.
$2n^3 + 8n^2 - 2n - 8 = 2n^2(n+4) - 2(n+4) = (2n^2 - 2)(n+4)$

Check: Check using the FOIL method.
$(2n^2 - 2)(n+4) = 2n^3 + 8n^2 - 2n - 8$

Factor the following polynomials by grouping. Be sure to arrange terms in descending order first.

1. $a^3 - 3 - 3a^2 + a$

2. $3c^2 - 4c + c^3 - 12$

3. $x^3 - 28 - 4x^2 + 7x$

4. $-8 + y^3 - y + 8y^2$

5. $b^3 - 15 - 5b^2 + 3b$

6. $d^3 + 20 - 4d - 5d^2$

7. $-3y^2 - 18 + y^3 + 6y$

8. $x^3 - 2x + 5x^2 - 10$

9. $-2y^2 - 3y + 6 + y^3$

10. $6a^2 - 3a - 18 + a^3$

11. $b^3 - 5 + b - 5b^2$

12. $c^3 - 14 - 7c + 2c^2$

13. $3d^2 - 4d - 12 + d^3$

14. $12 + a^3 + 6a + 2a^2$

15. $x^3 - 20 + 4x^2 - 5x$

16. $y^3 - 8y - 8 + y^2$

17. $b^3 - 6 - 3b^2 + 2b$

18. $-7 - c + c^3 + 7c^2$

6.2 Finding Roots of Polynomial Equations

A higher degree polynomial equation is one in which the largest exponent is greater than 2.

Fundamental Theorem of Algebra - Polynomials have the same number of roots as the value of its largest exponent.

Factor Theorem - A polynomial $f(x)$ has a factor $x - k$ if and only if $f(x) = 0$.

Remainder Theorem - If $f(x)$ is divided by $x - k$, then the remainder is $r = f(k)$.

To find the real and complex roots of a higher degree polynomial equation in standard form, first the **Rational Root Theorem** can be used to find all of the possible rational roots.

Next, the **Factor Theorem** can be used to test each of the possible rational roots and determine which ones are actual roots of the equation. If the equation has any roots that aren't rational, such as radical or complex roots, then polynomial division can often be used to produce a quadratic equation, at which point the quadratic formula can be used to find the non-rational roots. Also, if one radical root or one complex root of an equation is given, then another radical root or complex root of the equation can automatically be determined by finding the conjugate of the known root, or in other words, by switching the sign of the known root.

Example 2: Find all of the roots of the equation $x^3 - 2x^2 - 5x + 6 = 0$.

Step 1: Use the Rational Root Theorem to find all of the possible rational roots of the equation.

Since the equation is in standard form, all of its possible rational roots can be found by dividing all of the positive and negative factors of the last term on the left side by all of the positive and negative factors of the coefficient of the first term. Since the last term on the left side is 6, all of its positive and negative factors are ± 1, ± 2, ± 3, and ± 6. Also, since the coefficient of the first term is 1, all of its positive and negative factors are ± 1.

This means that all of the possible rational roots of the equation are $x = \frac{\pm 1}{\pm 1}$, $\frac{\pm 2}{\pm 1}$, $\frac{\pm 3}{\pm 1}$, and $\frac{\pm 6}{\pm 1}$, or $x = \pm 1$, ± 2, ± 3, and ± 6.

Step 2: Use the Factor Theorem to test each of the possible rational roots.

According to the Factor Theorem, if a root of the equation is plugged in for x, the left side of the equation becomes 0. With this in mind, all of the possible rational roots of the equation can be tested. First, $x = 1$ and $x = -1$ can be tested as follows.

$$f(x) = x^3 - 2x^2 - 5x + 6$$

$$f(1) = (1)^3 - 2(1)^2 - 5(1) + 6 = 1 - 2 - 5 + 6 = 0$$

$$f(-1) = (-1)^3 - 2(-1)^2 - 5(-1) + 6 = -1 - 2 + 5 + 6 = 8$$

This means that $x = 1$ is a root of the equation, but $x = -1$ is not.

Step 3:   Next, $x = 2$ and $x = -2$ can be tested as follows.

$$f(x) = x^3 - 2x^2 - 5x + 6$$

$$f(2) = (2)^3 - 2(2)^2 - 5(2) + 6 = 8 - 8 - 10 + 6 = -4$$

$$f(-2) = (-2)^3 - 2(-2)^2 = 5(-2) + 6 = -8 - 8 + 10 + 6 = 0$$

This means that $x = -2$ is a root of the equation, but $x = 2$ is not.

Step 4:   Next, $x = 3$ and $x = -3$ can be tested as follows.

$$f(x) = x^3 - 2x^2 - 5x + 6$$

$$f(3) = (3)^3 - 2(3)^2 - 5(3) + 6 = 27 - 18 - 15 + 6 = 0$$

$$f(-3) = (-3)^3 - 2(-3)^2 = 5(-3) + 6 = -27 - 18 + 15 + 6 = -24$$

This means that $x = 3$ is a root of the equation, but $x = -3$ is not. At this point, it has been determined that $x = 1, -2$, and 3 are roots of the equation $x^3 - 2x^2 - 5x + 6 = 0$. Since the highest exponent in the equation is 3, the Fundamental Theorem of Algebra states that the equation has 3 roots, which means that all of the roots have been found. Therefore, $x = 6$ and $x = -6$ do not have to be tested, because neither will be a root of the equation.

Find all of the roots of each of the following equations.

1. $x^3 - 3x^2 - 10x + 24 = 0$

2. $x^3 + 5x^2 - x - 5 = 0$

3. $x^3 + 4x^2 - 4x - 16 = 0$

4. $2x^3 - 5x^2 - 4x + 3 = 0$

5. $x^4 + 5x^3 + 5x^2 - 5x - 6 = 0$

6. $x^5 + x^4 - 13x^3 - 13x^2 + 36x + 36 = 0$

Example 3:   Find all of the roots of the equation $x^3 + x^2 - 2x - 2 = 0$.

Step 1:   Use the Rational Root Theorem to find all of the possible rational roots of the equation.

Since the equation is in standard form, all of its possible rational roots can be found by dividing all of the positive and negative factors of the last term on the left side by all of the positive and negative factors of the coefficient of the first term. Since the last term on the left side is -2, all of its positive and negative factors are ± 1 and ± 2. Also, since the coefficient of the first term is 1, all of its positive and negative factors are ± 1.

This means that all of the possible rational roots of the equation are

$$x = \frac{\pm 1}{\pm 1} \text{ and } \frac{\pm 2}{\pm 1}, \text{ or } x = \pm 1 \text{ and } \pm 2.$$

Step 2:   Use the Factor Theorem to test each of the possible rational roots.

According to the Factor Theorem, if a root of the equation is plugged in for x, the left side of the equation becomes 0. With this in mind, all of the possible rational roots of the equation can be tested. First, $x = 1$ and $x = -1$ can be tested as follows.

$$f(x) = x^3 + x^2 - 2x - 2$$
$$f(1) = (1)^3 + (1)^2 - 2(1) - 2 = 1 + 1 - 2 - 2 = -2$$
$$f(1) = (-1)^3 + (-1)^2 - 2(-1) - 2 = -1 + 1 + 2 - 2 = 0$$

This means that $x = -1$ is a root of the equation, but $x = 1$ is not.

Step 3:   Next, $x = 2$ and $x = -2$ can be tested as follows.

$$f(x) = x^3 + x^2 - 2x - 2$$
$$f(2) = (2)^3 + (2)^2 - 2(2) - 2 = 8 + 4 - 4 - 2 = 6$$
$$f(-2) = (-2)^3 + (-2)^2 - 2(-2) - 2 = -8 + 4 + 4 - 2 = -2$$

This means that neither $x = 2$ nor $x = -2$ is a root of the equation. At this point, there are no more possible rational roots to test, so the equation must have some roots that are not rational. Since the highest exponent in the equation is 3, the Fundamental Theorem of Algebra states that the equation has 3 roots, which means that there are two additional roots left to find.

Step 4:   Use polynomial division to produce a quadratic equation.

The two additional roots can be found by using polynomial division to produce a quadratic equation. Since $x = -1$ is a root of the equation $x^3 + x^2 - 2x - 2 = 0$, a factor of $x^3 + x^2 - 2x - 2$ must be $x + 1$, so $x^3 + x^2 - 2x - 2$ can be divided by $x + 1$ to produce one side of the quadratic equation. The division is done as follows.

$$
\begin{array}{r}
x^2 \qquad\quad -\ 2 \\
x+1 \overline{\smash{\big)}\ x^3\ +\ x^2\ -\ 2x\ -\ 2} \\
\underline{x^3\ +\ x^2} \qquad\qquad\quad \\
-\ 2x\ -\ 2 \\
\underline{-\ 2x\ -\ 2} \\
0
\end{array}
$$

Since $x^3 + x - 2x - 2$ divided by $x + 1$ equals $x^2 - 2$, the quadratic equation that is produced is $x^2 - 2 = 0$.

Step 5:   Use the quadratic formula to find the remaining roots.

Now that there is a quadratic equation in the form $ax^2 + bx + c = 0$, the quadratic formula can be used to solve it. The quadratic formula is

$$x = \frac{-b \pm \sqrt{b^2 - 4ac}}{2a},$$ and in this case, $a = 1$, $b = 0$, and $c = -2$. Thus, the

remaining roots can be found as follows.

$$x = \frac{-b \pm \sqrt{b^2 - 4ac}}{2a} = \frac{0 \pm \sqrt{0^2 - 4\,(1)\,(-2)}}{2\,(1)} = \frac{\pm\sqrt{8}}{2} = \frac{\pm 2\sqrt{2}}{2} = \pm\sqrt{2}$$

This means that the three roots of the equations $x^3 + x^2 - 2x - 2 = 0$ are $x = -1, -\sqrt{2},$ and $\sqrt{2}$.

Find all of the roots of each of the following equations.

1. $x^3 + 2x^2 - 3x - 6 = 0$

2. $x^3 - 3x^2 - 5x + 15 = 0$

3. $x^3 - x^2 - 6x + 6 = 0$

4. $x^3 - 2x^2 - 10x + 20 = 0$

5. $x^3 + 4x^2 - 8x - 32 = 0$

6. $x^4 - 8x^2 + 7 = 0$

Example 4:   Find all of the roots of the equation $x^4 + x^3 - 2x^2 + 4x - 24 = 0$. One of the roots is $x = 2i$.

Step 1:   Find the conjugate of the known root.

The root $x = 2i$ is a complex root, which means that another root of the equation can automatically be determined by finding the conjugate of the known root, or in other words, by switching the sign of the known root. This means that another root of the equation is $x = -2i$.

Step 2:   Use the Rational Root Theorem to find all of the possible rational roots of the equation.

Since the equation is in standard form, all of its possible rational roots can be found by dividing all of the positive and negative factors of the last term on the left side by all of the positive and negative factors of the coefficient of the first term. Since the last term on the left side is 24, all of its positive and negative factors are $\pm 1, \pm 2, \pm 3, \pm 4, \pm 6, \pm 8, \pm 12$, and ± 24. Also, since the coefficient of the first term is 1, all of its positive and negative factors are ± 1. This means that all of the possible rational roots of the equation are

$$x = \frac{\pm 1}{\pm 1}, \frac{\pm 2}{\pm 1}, \frac{\pm 3}{\pm 1}, \frac{\pm 4}{\pm 1}, \frac{\pm 6}{\pm 1}, \frac{\pm 8}{\pm 1}, \frac{\pm 12}{\pm 1}, \text{ and } \frac{\pm 24}{\pm 1},$$

or $x = \pm 1, \pm 2, \pm 3, \pm 4, \pm 6, \pm 8, \pm 12$, and ± 24.

Step 3:   Use the Factor Theorem to test each of the possible rational roots.

According to the Factor Theorem, if a root of the equation is plugged in for x, the left side of the equation becomes 0. With this in mind, all of the possible rational roots of the equation can be tested. First, $x = 1$ and $x = -1$ can be tested as follows.

$$f(x) = x^4 + x^3 - 2x^2 - 4x - 24$$

$$f(1) = (1)^4 + (1)^3 - 2(1)^2 - 4(1) - 24 = 1 + 1 - 2 + 4 - 24 = -20$$

$$f(-1) = (-1)^4 + (-1)^3 - 2(-1)^2 - 4(-1) - 24 = 1 - 1 - 2 - 4 - 24 = -30$$

This means that neither $x = 1$ nor $x = -1$ is a root of the equation.

Step 4:   Next, $x = 2$ and $x = -2$ can be tested as follows.

$$f(x) = x^4 + x^3 - 2x^2 - 4x - 24$$

$$f(2) = (2)^4 + (2)^3 - 2(2)^2 - 4(2) - 24 = 16 + 8 - 8 + 8 - 24 = 0$$

$$f(-2) = (-2)^4 + (-2)^3 - 2(-2)^2 - 4(-2) - 24 = 16 - 8 - 8 - 8 - 24 = -32$$

This means that $x = 2$ is a root of the equation, but $x = -2$ is not.

Step 5:   Next, $x = 3$ and $x = -3$ can be tested as follows.

$$f(x) = x^4 + x^3 - 2x^2 - 4x - 24$$

$$f(2) = (3)^4 + (3)^3 - 2(3)^2 - 4(3) - 24 = 81 + 27 - 18 + 12 - 24 = 78$$

$$f(-3) = (-3)^4 + (-3)^3 - 2(-3)^2 - 4(-3) - 24 = 81 - 27 - 18 - 12 - 24 = -0$$

This means that $x = -3$ is a root of the equation, but $x = 3$ is not. At this point, it has been determined that $x = 2i$, $-2i$, 2, and -3 are roots of the equation $x^4 + x^3 - 2x^2 + 4x - 24 = 0$. Since the highest exponent in the equation is 4, the Fundamental Theorem of Algebra states that the equation has 4 roots, which means that all of the roots have been found. Therefore, the rest of the possible rational roots do not have to be tested, because none will be a root of the equation.

Find all of the roots of each of the following equations. For each equation, one of the roots has been given.

1. $x^3 - x^2 + 9x - 9 = 0$; $x = 3i$

2. $x^3 + 15x^2 + x + 15 = 0$; $x = -i$

3. $x^4 + x^3 + 2x^2 + 4x - 8 = 0$; $x = 2i$

4. $x^4 + 21x^2 - 100 = 0$; $x = 5i$

5. $x^4 - x^3 + 34x^2 - 36x - 72 = 0$; $x = -6i$

6. $x^5 - 3x^4 + 15x^3 - 45x^2 - 16x + 48 = 0$; $x = 4i$

6.3 Solving Polynomial Equations Analytically

To solve a polynomial equation analytically, first it's necessary to write the equation in standard form, which means that the terms on the left side of the equation are in order from the highest degree to the lowest degree and the right side of the equation is 0. The equation can then often be solved by simply factoring, or if necessary, by using tools such as the Rational Root Theorem, the Factor Theorem, polynomial division, and the quadratic formula. According to the Fundamental Theorem of Algebra, a polynomial equation has the same number of solutions as its highest exponent, but the same solution can occur more than once. A solution that occurs m times is called a solution of multiplicity m.

Example 5: Solve the equation $2x + x^2 = 15$ for x.

Step 1: Write the equation in standard form.
To solve the equation $2x + x^2 = 15$ for x, first it's necessary to write the equation in standard form, which means that the terms on the left side of the equation are in order from the highest degree to the lowest degree and the right side of the equation is 0. This is done as follows.
$$2x + x^2 = 15$$
$$2x + x^2 - 15 = 15 - 15$$
$$2x + x^2 - 15 = 0$$
$$x^2 + 2x - 15 = 0$$

Step 2: Factor the left side of the equation.
Since the equation is quadratic, with the middle term on the left side having a coefficient of 2 and the last term being -15, the left side can be factored as follows.
$$x^2 + 2x - 15 = 0$$
$$(x + 5)(x - 3) = 0$$

Step 3: Solve for x.
Since $(x + 5)(x - 3) = 0$, either $x + 5 = 0$ or $x - 3 = 0$. This means that x can be solved for as follows.
$$x + 5 = 0$$
$$x + 5 - 5 = 0 - 5$$
$$x = -5$$
or
$$x - 3 = 0$$
$$x - 3 + 3 = 0 + 3$$
$$x = 3$$
Therefore, the solution to the equation $2x + x^2 = 15$ is $x = -5$ and 3.

Solve each of these equations for x.

1. $x^2 = 4x + 12$
2. $x^2 - 2x = 35$
3. $x + x^2 = 72$
4. $2x^2 + 29x = 15$
5. $12x + 4 = -9x^2$
6. $25x^2 = 144$

Example 6:   Solve the equation $x^3 + 27x = 40 - 12x^2$ for x.

Step 1:   Write the equation in standard form.

To solve the equation $x^3 + 27x = 40 - 12x^2$ for x, first it's necessary to write the equation in standard form, which means that the terms on the left side of the equation are in order from the highest degree to the lowest degree and the right side of the equation is 0. This is done as follows.

$$x^3 + 27x = 40 - 12x^2$$
$$x^3 + 27x - 40 + 12x^2 = 40 - 12x^2 - 40 + 12x^2$$
$$x^3 + 27x - 40 + 12x^2 = 0$$
$$x^3 + 12x^2 + 27x - 40 = 0$$

Step 2:   Use the Rational Root Theorem to find all of the possible rational roots of the equation.

Since the equation is now in standard form, all of its possible rational roots can be found by dividing all of the positive and negative factors of the last term on the left side by all of the positive and negative factors of the coefficient of the first term. Since the last term on the left side is 40, all of its positive and negative factors are $\pm 1, \pm 2, \pm 4, \pm 5, \pm 8, \pm 10, \pm 20$, and ± 40. Also, since the coefficient of the first term is 1, all of its positive and negative factors are ± 1. This means that all of the possible rational roots of the equation are

$$x = \frac{\pm 1}{\pm 1}, \frac{\pm 2}{\pm 1}, \frac{\pm 4}{\pm 1}, \frac{\pm 5}{\pm 1}, \frac{\pm 8}{\pm 1}, \frac{\pm 10}{\pm 1}, \frac{\pm 20}{\pm 1}, \text{ and } \frac{\pm 40}{\pm 1}, \text{ or}$$

$$x = \pm 1, \pm 2, \pm 4, \pm 5, \pm 8, \pm 10, \pm 20, \text{ and } \pm 40.$$

Step 3:   Use the Factor Theorem to find a rational root.

According to the Factor Theorem, if a root of the equation is plugged in for x, the left side of the equation becomes 0. With this in mind, all of the possible rational roots of the equation can be tested. First, $x = 1$ and $x = -1$ can be tested as follows.

$$f(x) = x^3 + 12x^2 + 27x - 40$$
$$f(1) = (1)^3 + 12(1)^2 + 27(1) - 40 = 1 + 12 + 27 - 40 = 0$$
$$f(-1) = (-1)^3 + 12(-1)^2 + 27(-1) - 40 = -1 + 12 - 27 - 40 = -56$$

This means that $x = 1$ is a solution to the equation, but $x = -1$ is not. At this point, only one of the solutions to the equation has been found, but the Fundamental Theorem of Algebra states that the equation has 3 solutions, since the highest exponent on the left side of the equation is 3. While it's possible to keep using the Factor Theorem to test the other possible rational roots, a faster approach may be to produce a quadratic equation to find the two additional solutions.

Step 4:   Use polynomial division to produce a quadratic equation.

Since $x = 1$ is a solution to the equation $x^3 + 12x^2 + 27x - 40 = 0$, a factor of $x^3 + 12x^2 + 27x - 40$ must be $x - 1$, so $x^3 + 12x^2 + 27x - 40$ can be divided by $x - 1$ to produce one side of the quadratic equation. The division is done synthetically as follows.

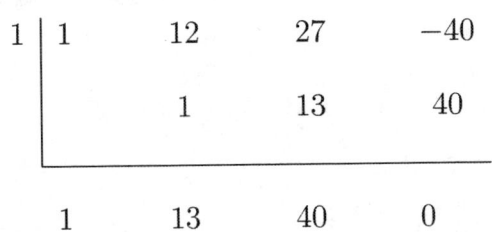

Since $x^3 + 12x^2 + 27x - 40$ divided by $x - 1$ equals $x^2 + 13x + 40$, the quadratic equation that is produced is $x^2 + 13x + 40 = 0$.

Step 5:   Factor the left side of the quadratic equation.

Since the middle term on the left side of the quadratic equation has a coefficient of 13 and the last term is 40, the left side can be factored as follows.

$$x^2 + 13x + 40 = 0$$
$$(x + 5)(x + 8) = 0$$

Step 6:   Solve for x.

Since $(x + 5)(x + 8) = 0$, either $x + 5 = 0$ or $x + 8 = 0$. This means that x can be solved for as follows.

$$x + 5 = 0$$
$$x + 5 - 5 = 0 - 5$$
$$x = -5$$

or

$$x + 8 = 0$$
$$x + 8 - 8 = 0 - 8$$
$$x = -8$$

Therefore, the solution to the equation $x^3 + 27x = 40 - 12x^2$ is $x = -8, -5$, and 1.

Solve each of these equations for x.

1. $x^3 + 9x^2 = -24 - 26x$

2. $x^3 + 24 = 6x^2 + 4x$

3. $x^3 + 2x^2 = 55x + 56$

4. $x^3 + 45 = 5x^2 + 41x$

5. $x^3 + 47x = 12x^2 + 60$

6. $11x^2 + 16x = 84 - x^3$

Example 7:   Solve the equation $x^4 + 2x^3 + 2x^2 + 2x = -1$ for x. Include both real and complex solutions.

Step 1:   Write the equation in standard form.

To solve the equation $x^4 + 2x^3 + 2x^2 + 2x = -1$ for x, first it's necessary to write the equation in standard form, which means that the terms on the left side of the equation are in order from the highest degree to the lowest degree and the right side of the equation is 0. This is done as follows.

$$x^4 + 2x^3 + 2x^2 + 2x = -1$$

$$x^4 + 2x^3 + 2x^2 + 2x + 1 = -1 + 1$$

$$x^4 + 2x^3 + 2x^2 + 2x + 1 = 0$$

Step 2:   Use the Rational Root Theorem to find all of the possible rational roots of the equation.

Since the equation is now in standard form, all of its possible rational roots can be found by dividing all of the positive and negative factors of the last term on the left side by all of the positive and negative factors of the coefficient of the first term. Since the last term on the left side is 1, all of its positive and negative factors are ± 1. Also, since the coefficient of the first term is 1, all of its positive and negative factors are ± 1. This means that all of the possible rational roots of the equation are $x = \dfrac{\pm 1}{\pm 1}$, or $x = \pm 1$.

Step 3:   Use the Factor Theorem to find a rational root.

According to the Factor Theorem, if a root of the equation is plugged in for x, the left side of the equation becomes 0. With this in mind, $x = 1$ and $x = -1$ can be tested as follows.

$$f(x) = x^4 + 2x^3 + 2x + 1$$

$$f(1) = (1)^4 + 2(1)^3 + 2(1) + 1 = 1 + 2 + 2 + 2 + 1 = 8$$

$$f(-1) = (-1)^4 + 2(-1)^3 + 2(-1) + 1 = 1 - 2 + 2 - 2 + 1 = 0$$

This means that $x = -1$ is a solution to the equation, but $x = 1$ is not. At this point, only one of the solutions to the equation has been found, but the Fundamental Theorem of Algebra states that the equation has 4 solutions, since the highest exponent on the left side of the equation is 4. However, there are no additional possible rational roots to test, so the only option is to use polynomial division to produce a cubic equation.

Step 4:   Use polynomial division to produce a cubic equation.

Since $x = -1$ is a solution to the equation $x^4 + 2x^3 + 2x^2 + 2x + 1 = 0$, a factor of $x^4 + 2x^3 + 2x^2 + 2x + 1$ must be $x + 1$, so $x^4 + 2x^3 + 2x^2 + 2x + 1$ can be divided by $x + 1$ to produce one side of the cubic equation. The division is done synthetically as follows.

$$
\begin{array}{r|rrrrr}
-1 & 1 & 2 & 2 & 2 & 1 \\
   &   & -1 & -1 & -1 & -1 \\
\hline
   & 1 & 1 & 1 & 1 & 0
\end{array}
$$

Since $x^4 + 2x^3 + 2x^2 + 2x + 1$ divided by $x + 1$ equals $x^3 + x^2 + x + 1$, the cubic equation that is produced is $x^3 + x^2 + x + 1 = 0$.

Step 5:   Use the Factor Theorem to check if the rational root already found occurs more than once.

Even though it has already been determined that $x = -1$ is a solution to the equation $x^4 + 2x^3 + 2x^2 + 2x + 1 = 0$, it may be a solution more than once. To find out if this is the case, the root $x = -1$ can be tested again with the cubic equation $x^3 + x^2 + x + 1 = 0$ as follows.

$$f(x) = x^3 + x^2 + x + 1$$

$$f(-1) = (-1)^3 + (-1)^2 + (-1) + 1 = -1 + 1 + (-1) + 1 = 0$$

This means that $x = -1$ is a solution to the equation $x^4 + 2x^3 + 2x^2 + 2x + 1 = 0$ twice, or in other words, that it is a solution of multiplicity 2. However, again there are no additional possible rational roots to test, so the only option is to use polynomial division to produce a quadratic equation.

Step 6:   Use polynomial division to produce a quadratic equation.

Since $x = -1$ is a solution to the equation $x^3 + x^2 + x + 1 = 0$, a factor of $x^3 + x^2 + x + 1$ must be $x + 1$, so $x^3 + x^2 + x + 1$ can be divided by $x + 1$ to produce one side of the quadratic equation. The division is done synthetically as follows.

$$
\begin{array}{r|rrrr}
-1 & 1 & 1 & 1 & 1 \\
   &   & -1 & 0 & -1 \\
\hline
   & 1 & 0 & 1 & 0
\end{array}
$$

Since $x^3 + x^2 + x + 1$ divided by $x + 1$ equals $x^2 + 1$, the quadratic equation that is produced is $x^2 + 1 = 0$.

Step 7:   Use the quadratic formula to find the remaining roots.

Now that there is a quadratic equation in the form $ax^2 + bx + c = 0$, the quadratic formula can be used to solve it. The quadratic formula is

$$x = \frac{-b \pm \sqrt{b^2 - 4ac}}{2a}$$, and in this case $a = 1$, $b = 0$, and $c = 1$. Thus, the

remaining roots can be found as follows.

$$x = \frac{-b \pm \sqrt{b^2 - 4ac}}{2a} = \frac{0 \pm \sqrt{0^2 - 4\,(1)\,(1)}}{2\,(1)} =$$

$$\frac{\pm\sqrt{-4}}{2} = \frac{\pm i\sqrt{4}}{2} = \frac{\pm 2i}{2} = \pm i$$

This means that the solution to the equation $x^4 + 2x^3 + 2x^2 + 2x = -1$ is $x = -1, -i$, or i.

Solve each of these equations for x. Include both real and complex solutions.

1. $x^3 + x^2 - 8x = 12$

2. $x^3 - 5x^2 + 48 = 8x$

3. $x^4 + 5x^2 + 4 = 2x^3 + 8x$

4. $x^4 + 20x = 4x^3 + x^2 + 20$

5. $x^4 + 10x^3 + 33x^2 + 40x = -16$

6. $x^4 + 18x^2 + 81 = 6x^3 + 54x$

6.4 Solving Polynomial Equations Graphically

There are two ways to solve a polynomial equation graphically.

1) Graph the left side of the equation, graph the right side of the equation on the same coordinate grid, and determine the x-coordinates of the points where the graphs intersect.

2) Write the polynomial equation in standard form, graph the left side of the equation, and determine the x-coordinates of the points where the graph crosses the x-axis.

If the first approach is used and the graphs do not intersect, or if the second approach is used and the graph does not cross the x-axis, then the polynomial equation has no real solution. When solving polynomial equations graphically, only real solutions, and not complex solutions, can be found.

Example 8: Solve the equation $x^3 + 3x^2 = x + 3$ graphically.

Step 1: Graph the left side of the equation.
First, the function $f(x) = x^3 + 3x^2$ should be graphed as follows.

> **Calculator:**
> On the TI-83/84 graphing calculator, this graph can be produced by performing the following steps.
>
> a. Press $\boxed{Y=}$.
>
> b. Enter X^3 + 3X^2 after $Y_1 =$.
>
> c. Press $\boxed{\text{WINDOW}}$.
>
> d. Enter the following values.
>
> Xmin = –6
> Xmax = 4
> Xscl = 1
> Ymin = –5
> Ymax = 5
> Yscl = 1
> Xres = 1
>
> e. Press $\boxed{\text{GRAPH}}$.

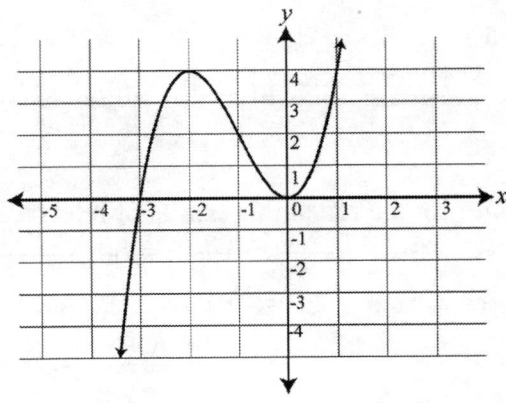

Step 2:   Graph the right side of the equation on the same coordinate grid.

Next, the function $g(x) = x + 3$ should be graphed on the same coordinate grid as follows.

Calculator:

On the TI-83/84 graphing calculator, this graph can be produced by performing the following steps.

a. Press $\boxed{Y=}$.

b. Enter X + 3 after Y$_2$ =.

c. Press $\boxed{\text{WINDOW}}$.

d. Enter the following values.

Xmin = –6
Xmax = 4
Xscl = 1
Ymin = –5
Ymax = 5
Yscl = 1
Xres = 1

e. Press $\boxed{\text{GRAPH}}$.

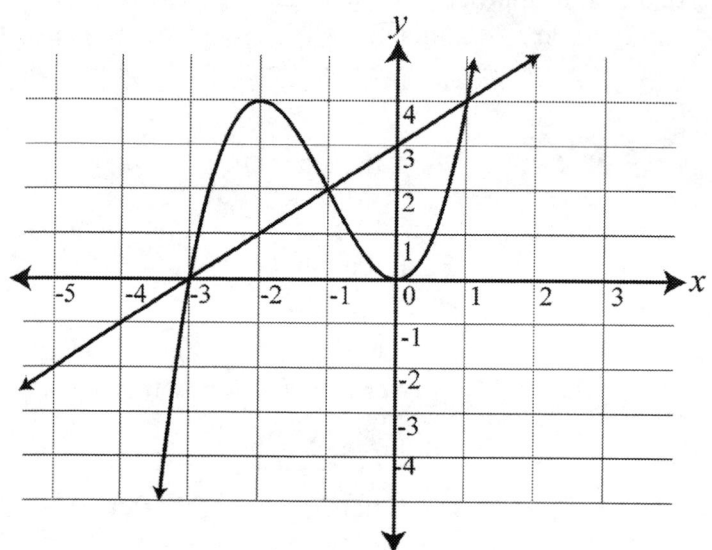

Step 3:   Determine the x-coordinates of the points where the graphs intersect.

Calculator:

On the TI-83/84 graphing calculator, the points of intersection of the graphs can be found by performing the following steps.

a. Press $\boxed{\text{2ND}}$ $\boxed{\text{TRACE}}$.

b. Press $\boxed{5}$.

c. Press $\boxed{\text{ENTER}}$ twice.

d. Move the cursor as close as possible to a point of intersection.

e. Press $\boxed{\text{ENTER}}$.

f. Repeat the steps above for the remaining points of intersection.

Since the graphs intersect at the points $(-3, 0)$, $(-1, 2)$, and $(1, 4)$, the solution to the equation $x^3 + 3x^2 = x + 3$ is $x = -3, -1$, and 1.

Example 9: Solve the equation $x^4 + 5x^2 + 24x = 6x^3 + 36$ graphically.

Step 1: Write the equation in standard form.

To write the equation $x^4 + 5x^2 + 24x = 6x^3 + 36$ in standard form, the terms on the left side of the equation should be in order from the highest degree to the lowest degree and the right side of the equation should be 0. The equation can be converted to standard form as follows.

$$x^4 + 5x^2 + 24x = 6x^3 + 36$$

$$x^4 + 5x^2 + 24x - 6x^3 - 36 = 6x^3 + 36 - 6x^3 - 36$$

$$x^4 + 5x^2 + 24x - 6x^3 - 36 = 0$$

$$x^4 - 6x^3 + 5x^2 + 24x - 36 = 0$$

Step 2: Graph the left side of the equation.

Now, the function $f(x) = x^4 - 6x^3 + 5x^2 + 24x - 36$ should be graphed as follows.

Calculator:
On the TI-83/84 graphing calculator, this graph can be produced by performing the following steps.

a. Press $\boxed{Y=}$.

b. Enter X^4 – 6X^3 + 5X^2 + 24X – 36 after $Y_1 =$.

c. Press $\boxed{\text{WINDOW}}$.

d. Enter the following values.

Xmin = –4
Xmax = 6
Xscl = 1
Ymin = –55
Ymax = 15
Yscl = 15
Xres = 1

e. Press $\boxed{\text{GRAPH}}$.

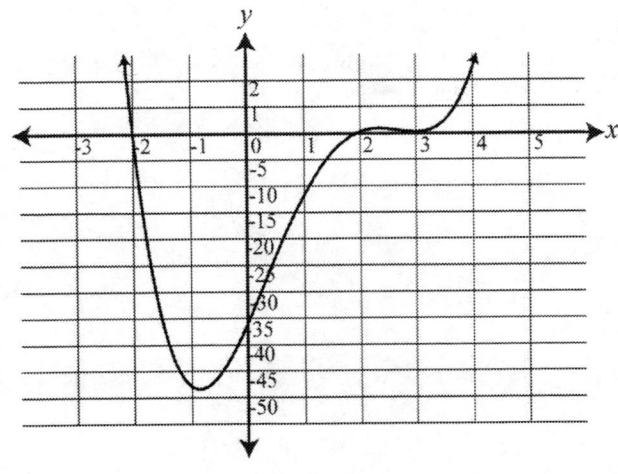

Step 3:   Determine the x-coordinates of the points where the graph crosses the x-axis.

> **Calculator:**
> On the TI-83/84 graphing calculator, the points where the graph crosses the x-axis can be found by performing the following steps.
>
> a. Press ⎡2ND⎤ ⎡TRACE⎤.
>
> b. Press ⎡2⎤.
>
> c. Move the cursor to the left of a point where the graph crosses the x-axis.
>
> d. Press ⎡ENTER⎤.
>
> e. Move the cursor to the right of the point where the graph crosses the x-axis.
>
> f. Press ⎡ENTER⎤ twice.
>
> g. Repeat the steps above for the remaining points where the graph crosses the x-axis.

Since the graph crosses the x-axis at the points $(-2, 0)$, $(2, 0)$, and $(3, 0)$, the solution to the equation $x^4 + 5x^2 + 24x = 6x^3 + 36$ is $x = -2, 2$, and 3.

Solve each of the following equations graphically.

1. $x^3 + x^2 = 9x + 9$

2. $x^3 + 32 = 16x + 2x^2$

3. $x^3 + 11x^2 + 23x = 35$

4. $6x^2 + x^3 = -12x - 8$

5. $44x + x^3 = 48 + 12x^2$

6. $2x^3 + 15x^2 + 24x = 16$

7. $x^3 + 8x^2 = 15x + 54$

8. $x^3 = 21x + 20$

9. $x^4 + x^3 + 50 = 27x^2 + 25x$

10. $x^4 + 2x^3 = 35x^2 + 72x + 36$

11. $x^4 + 5x + 14 = 5x^3 + 15x^2$

12. $x^5 + 4x + 10x^2 = 5x^3 + 2x^4 + 8$

Example 10:   How many distinct real solutions does the equation $5x^2 + x^4 = -6$ have?

Step 1:   Write the equation in standard form.

To write the equation $5x^2 + x^4 = -6$ in standard form, the terms on the left side of the equation should be in order from the highest degree to the lowest degree and the right side of the equation should be 0. The equation can be converted to standard form as follows.

$$5x^2 + x^4 = -6 \rightarrow 5x^2 + x^4 + 6 = 0 \rightarrow x^4 + 5x^2 + 6 = 0$$

Step 2:   Graph the left side of the equation.

Now, the function $f(x) = x^4 + 5x^2 + 6$ should be graphed as follows.

Calculator:
On the TI-83/84 graphing calculator, this graph can be produced by performing the following steps.

a. Press $\boxed{Y=}$.

b. Enter X^4 + 5X^2 + 6 after $Y_1 =$.

c. Press $\boxed{\text{WINDOW}}$.

d. Enter the following values.

Xmin = –5
Xmax = 5
Xscl = 1
Ymin = –10
Ymax = 40
Yscl = 5
Xres = 1

e. Press $\boxed{\text{GRAPH}}$.

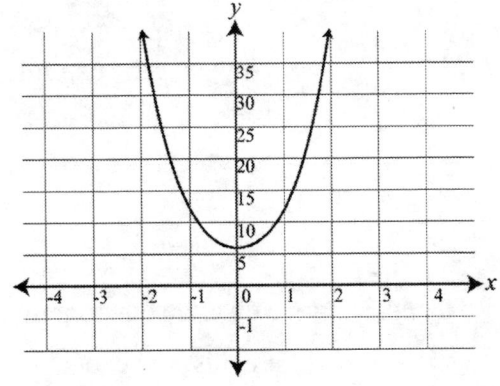

Step 3:   Count the number of times the graph crosses the x-axis.

The number of distinct real solutions the equation has is equal to the number of times its graph crosses the x-axis. Since the graph does not cross the x-axis, the equation has 0 distinct real solutions.

Determine the number of distinct real solutions to each of the following equations.

1. $6x^2 + x^4 = -5$

2. $x^3 + 4x = 28 + 7x^2$

3. $x^4 = 4x^2 + 45$

4. $3x^3 + x^5 + 2x = 9x^2 + 3x^4 + 6$

5. $x^6 + 27x^4 + 51x^2 = -25$

6. $x^6 + 8x^4 = -9x^2 + 18$

6.5   Solving Exponential Equations Analytically

To solve an exponential equation analytically, the base and exponent must first be isolated on one side of the equation. Once the base and exponent are isolated, the logarithm of both sides of the equation should be taken. The base of the logarithm doesn't matter, but often, it's the natural logarithm that's taken. At this point, the property $\log a(c^d) = d \log_a c$ can be used, if necessary, so that it's possible to solve for the variable in the exponent.

Example 11:   Solve the equation $4^x = 125$ for x. Round your answer to two decimal places.

Step 1:   Take the natural logarithm of both sides of the equation.

Since the base and exponent are already isolated on one side of the equation, the natural logarithm of both sides of the equation should be taken as follows.

$$4^x = 125$$

$$\ln(4^x) = \ln(125)$$

Step 2:   Use the property $\log_a(c^d) = d \log_a c$.

Now that the natural logarithm of both sides of the equation has been taken, the property $\log_a(c^d) = d \log_a c$ can be used as follows.

$$\ln(4^x) = \ln(125)$$

$$x \cdot \ln(4) = \ln(125)$$

Step 3:   Solve the equation for x.

Finally, the equation can be solved for x as follows.

$$x \cdot \ln(4) = \ln(125)$$

$$\frac{x \cdot \ln(4)}{\ln(4)} = \frac{\ln(125)}{\ln(4)}$$

$$x = \frac{\ln(125)}{\ln(4)}$$

$$x = \frac{4.8283}{1.3863}$$

Solve each of the following equations for x. Round your answers to two decimal places.

1. $3^x = 62$

2. $8^x = 350$

3. $\left(\frac{1}{5}\right)^x = 84$

4. $16^x = 48$

5. $7^x = 975$

6. $\left(\frac{1}{2}\right)^x = 600$

Example 12: Solve the equation $(4)\,(5^{x+2}) - 28 = 880$ for x. Round your answer to two decimal places.

Step 1: Isolate the base and exponent.

To isolate the base and exponent on one side of the equation, first 28 must be added to both sides of the equation and then both sides of the equation must be divided by 4 as follows.

$(4)\,(5^{x+2}) - 28 + 28 = 880 + 28$

$(4)\,(5^{x+2}) = 908$

$$\frac{(4)\,(5^{x+2})}{4} = \frac{908}{4}$$

$5^{x+2} = 227$

Step 2: Take the natural logarithm of both sides of the equation.

Since the base and exponent are now isolated on one side of the equation, the natural logarithm of both sides of the equation should be taken as follows.

$5^{x+2} = 227$

$\ln\left(5^{x+2}\right) = \ln\left(227\right)$

Step 3: Use the property $\log_a\left(c^d\right) = d\log_a c$.

Now that the natural logarithm of both sides of the equation has been taken, the property $\log_a\left(c^d\right) = d\log_a c$ can be used as follows.

$\ln\left(5^{x+2}\right) = \ln\left(227\right)$

$(x+2) \cdot \ln\left(5\right) = \ln\left(227\right)$

Step 4: Solve the equation for x.

Finally, the equation can be solved for x as follows.

$(x+2) \cdot \ln\left(5\right) = \ln\left(227\right)$

$$\frac{(x+2) \cdot \ln\left(5\right)}{\ln\left(5\right)} = \frac{\ln\left(227\right)}{\ln\left(5\right)}$$

$$x + 2 = \frac{\ln\left(227\right)}{\ln\left(5\right)}$$

$$x + 2 - 2 = \frac{\ln\left(227\right)}{\ln\left(5\right)} - 2$$

$$x = \frac{\ln\left(227\right)}{\ln\left(5\right)} - 2$$

$$x = \frac{5.4250}{1.6094} - 2 = 3.37 - 2 = 1.37$$

Solve each of the following equations for x. Round your answers to two decimal places.

1. $(2)(9^{x-1}) + 39 = 111$

2. $(5)(6^{x+2}) - 94 = 91$

3. $\left(\frac{3}{4}\right)(4^{x+7}) - 23 = 67$

4. $(5)(12^{2x-3}) + 19 = 784$

5. $(9)(7^{3x+5}) - 123 = 957$

6. $\left(\frac{1}{6}\right)(5^{8x-9}) + 44 = 170$

Example 13:   Solve the equation $10e^{x-3} + 37 = 447$ for x. Round your answer to two decimal places.

Step 1:   Isolate the base and exponent.

To isolate the base and exponent on one side of the equation, first 37 must be subtracted from both sides of the equation and then both sides of the equation must be divided by 10 as follows.

$$10e^{x-3} + 37 = 447$$
$$10e^{x-3} + 37 - 37 = 447 - 37$$
$$10e^{x-3} = 410$$

$$\frac{10e^{x-3}}{10} = \frac{410}{10}$$

$$e^{x-3} = 41$$

Step 2:   Take the natural logarithm of both sides of the equation.

Since the base and exponent are now isolated on one side of the equation, the natural logarithm of both sides of the equation should be taken as follows.

$$e^{x-3} = 41$$
$$\ln\left(e^{x-3}\right) = \ln\left(41\right)$$
$$x - 3 = \ln\left(41\right)$$

Step 3:   Solve the equation for x.

Finally, the equation can be solved for x as follows.

$$x - 3 = \ln\left(41\right)$$
$$x - 3 + 3 = \ln\left(41\right) + 3$$
$$x = \ln\left(41\right) + 3$$
$$x = 3.71 + 3 = 6.71$$

Solve each of the following equations for x. Round your answers to two decimal places.

7. $4e^{x+9} - 1 = 35$

8. $\dfrac{e^{x-5}}{8} + 56 = 202$

9. $25e^{10x-3} + 169 = 394$

10. $11e^{\frac{x}{4}} - 21 = 100$

11. $\dfrac{e^{2x+7}}{18} + 53 = 322$

12. $6e^{\frac{x-5}{6}} - 23 = 643$

6.6   Solving Exponential Equations Graphically

There are two ways to solve an exponential equation graphically.

1) Graph the left side of the equation, graph the right side of the equation on the same coordinate grid, and determine the x-coordinate of the point where the graphs intersect.

2) Move all the terms of the exponential equation to the left side so that the right side is 0, graph the left side, and determine the x-coordinate of the point where the graph crosses the x-axis.

If the first approach is used and the graphs do not intersect, or if the second approach is used and the graph does not cross the x-axis, then the exponential equation has no solution.

Example 14:   Solve the equation $(4)(3^x) - 28 = 8$ graphically.

Step 1:   Graph the left side of the equation.

First, the function $f(x) = (4)(3^x) - 28 = 8$ should be graphed as follows.

Calculator:
On the TI-83/84 graphing calculator, this graph can be produced by performing the following steps.

a. Press $\boxed{Y=}$.

b. Enter 4*3^X − 28 after $Y_1 =$.

c. Press $\boxed{WINDOW}$.

d. Enter the following values.

Xmin = −5
Xmax = 5
Xscl = 1
Ymin = −40
Ymax = 60
Yscl = 10
Xres = 1

e. Press $\boxed{GRAPH}$.

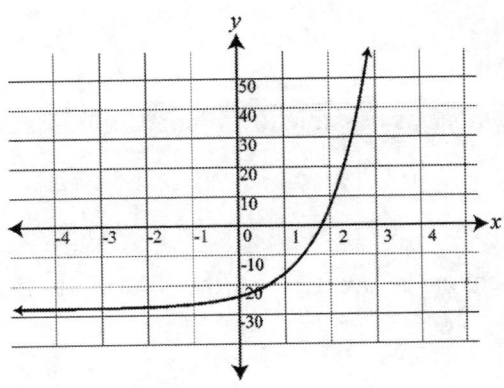

Step 2:   Graph the right side of the equation on the same coordinate grid.

Next, the function $g(x) = 8$ should be graphed on the same coordinate grid as follows.

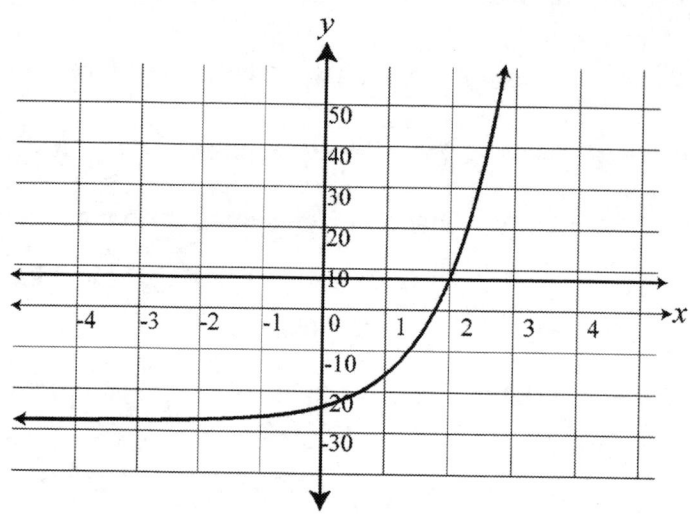

Calculator:
On the TI-83/84 graphing calculator, this graph can be produced by performing the following steps.

a. Press $\boxed{Y=}$.

b. Enter 8 after Y_2.

c. Press $\boxed{WINDOW}$.

d. Enter the following values.

Xmin = −5
Xmax = 5
Xscl = 1
Ymin = −40
Ymax = 60
Yscl = 10
Xres = 1

e. Press $\boxed{GRAPH}$.

Step 3:   Determine the x-coordinate of the point where the graphs intersect.

Calculator:
On the TI-83/84 graphing calculator, the points of intersection of the graphs can be found by performing the following steps.

a. Press $\boxed{2ND}$ $\boxed{TRACE}$.

b. Press $\boxed{5}$.

c. Press $\boxed{ENTER}$ twice.

d. Move the cursor as close as possible to a point of intersection.

e. Press $\boxed{ENTER}$.

Since the graphs intersect at the point $(2, 8)$, the solution to the equation $(4)(3^x) - 28 = 8$ is $x = 2$.

Example 15:   Solve the equation $(128)(8^x) + 9 = 11$ graphically.

Step 1:   Move all the terms to the left side of the equation.
When all the terms of $(128)(8^x) + 9 = 11$ are moved to the left side of the equation, the right side of the equation becomes 0 as follows.
$(128)(8^x) + 9 = 11 \rightarrow (128)(8^x) - 2 = 0$

Step 2:   Graph the left side of the equation.
Now, the function $f(x) = (128)(8^x) - 2$ should be graphed as follows.

Calculator:
On the TI-83/84 graphing calculator, this graph can be produced by performing the following steps.

a. Press $\boxed{Y =}$.

b. Enter $128*8\wedge X - 2$ after $Y_1 = .$

c. Press $\boxed{\text{WINDOW}}$.

d. Enter the following values.

Xmin = −7
Xmax = 3
Xscl = 1
Ymin = −20
Ymax = 80
Yscl = 10
Xres = 1

e. Press $\boxed{\text{GRAPH}}$.

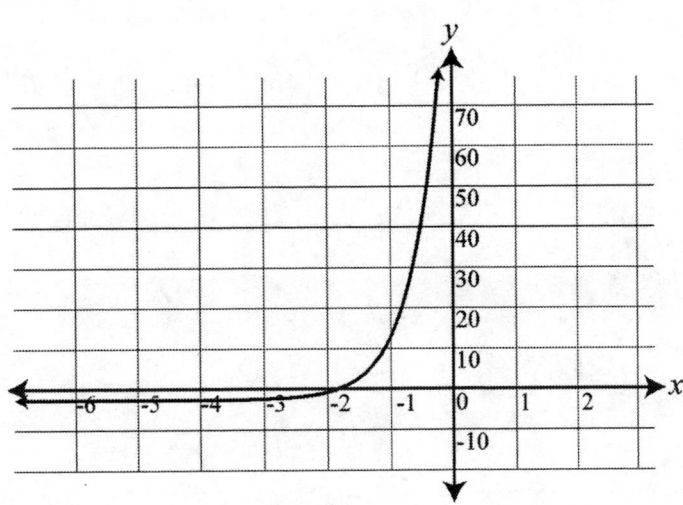

Step 3:   Determine the x-coordinate of the point where the graph crosses the x-axis.

Calculator:
On the TI-83/84 graphing calculator, the points of intersection of the graphs can be found by performing the following steps.

a. Press $\boxed{\text{2ND}}$ $\boxed{\text{TRACE}}$.

b. Press $\boxed{2}$.

c. Press $\boxed{\text{ENTER}}$.

d. Move the cursor as close as possible to a point of intersection.

e. Press $\boxed{\text{ENTER}}$ twice.

Since the graph crosses the x-axis at the point $(-2, 0)$, the solution to the equation $(128)(8^x) + 9 = 11$ is $x = -2$.

Solve each of the following equations graphically.

1. $(5)(2^x) + 17 = 57$

2. $(-4)(18^x) + 46 = -26$

3. $(12)(4^x) - 33 = -30$

4. $(98)(7^x) + 4 = 6$

5. $\left(\frac{1}{9}\right)(6^x) + 68 = 72$

6. $(7)(5^{x-4}) - 6 = 169$

7. $\left(\frac{1}{3}\right)(9^x) - 14 = 13$

8. $(42)(3^x) + 49 = 63$

9. $(-7)(2^x) + 16 = -208$

10. $(1,000)(5^x) - 94 = -86$

11. $(6)(7^{x+8}) + 91 = 385$

12. $\left(\frac{1}{8}\right)(4^{2x-1}) + 12 = 20$

Example 16: Determine is the equation $(3)(5^{x-2}) + 17 = 2$ has a solution.

Step 1: Move all the terms to the left side of the equation.
When all the terms of $(3)(5^{x-2}) + 17 = 2$ are moved to the left side of the equation, the right side of the equation becomes 0 as follows.
$(3)(5^{x-2}) + 17 = 2 \rightarrow (3)(5^{x-2}) + 15 = 0$

Step 2: Graph the left side of the equation.

Now, the function $f(x) = (3)(5^{x-2}) + 15 = 0$ should be graphed as follows.

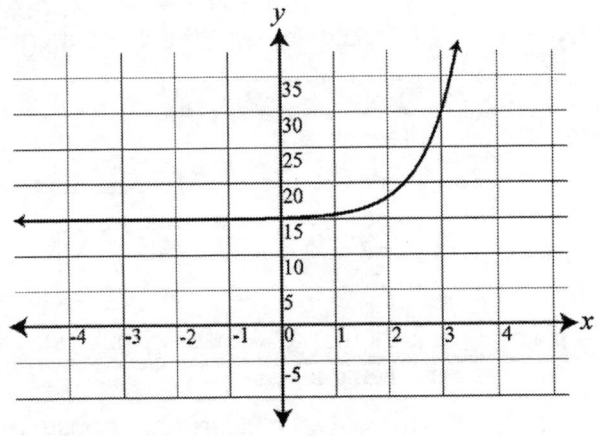

Calculator:
On the TI-83/84 graphing calculator, this graph can be produced by performing the following steps.

a. Press $\boxed{Y=}$.

b. Enter 3*5^(X − 2) + 15 after $Y_1 =$.

c. Press $\boxed{\text{WINDOW}}$.

d. Enter the following values.

Xmin = −5
Xmax = 5
Xscl = 1
Ymin = −10
Ymax = 40
Yscl = 5
Xres = 1

e. Press $\boxed{\text{GRAPH}}$.

Step 3: Determine if the graph crosses the x-axis.

If the graph crosses the x-axis, it has a solution, and if it doesn't cross the x-axis, it doesn't have a solution. Since the graph doesn't cross the x-axis, it doesn't have a solution.

Determine if each of the following equations has a solution.

1. $(2)(16^x) + 41 = 9$

2. $(-3)(11^x) + 400 = 37$

3. $\left(-\frac{1}{2}\right)(6^x) - 50 = 58$

4. $(80)(4^{x+7}) - 101 = -96$

5. $(8)\left(7^{\frac{x-3}{2}}\right) - 202 = 190$

6. $\left(\frac{1}{50}\right)(10^{5x-1}) + 66 = 46$

6.7   Solving Logarithmic Equations Analytically

Logarithmic equations can be solved analytically using various methods. One method is to use the fact that another way of writing $\log_a x = y$ is $x = a^y$. Other methods include using properties of logarithms, such as $\log_a(cd) = \log_a c + \log_a d$, $\log_a \left(\frac{c}{d}\right) = \log_a c - \log_a d$, and $\log_a \left(c^d\right) = d \log_a c$, or using the property that says if $\log_a c = \log_a d$, then $c = d$. In addition, to solve logarithmic equations analytically, it is often necessary to use algebra to find the values of unknown variables, and the values of the variables must be checked to make sure they are valid solutions.

Example 17:   Solve the equation $\log_6 \left(x^2 + x + 6\right) = 2$ for x.

Step 1:   Use the fact that another way of writing $\log_a x = y$ is $x = a^y$.

Since another way of writing $\log_a x = y$ is $x = a^y$, the equation can also be written as $x^2 + x + 6 = 6^2$, or $x^2 + x + 6 = 36$.

Step 2:   Write the equation $x^2 + x + 6 = 36$ in standard form.

$x^2 + x - 30 = 0$

Step 3:   Factor the left side of the equation.

$(x + 6)(x - 5) = 0$

Step 4:   Solve for x.

Since $(x + 6)(x - 5) = 0$, either $x + 6 = 0$ or $x - 5 = 0$. This means that x can be solved for as follows.

$x + 6 = 0 \qquad x - 5 = 0$
$x = -6 \qquad x = 5$

Step 5:   Check the values of x to make sure that only the logarithm of positive numbers are being taken.

Since only the logarithm of positive numbers can be taken, the values of x must be plugged into the original equation to make sure that they are valid solutions. The value -6 can be plugged in as follows.

$\log_6 \left(x^2 + x + 6\right) = 2$

$\log_6 \left((-6)^2 + (-6) + 6\right) = 2$

$\log_6 (36 - 6 + 6) = 2$

$\log_6 (36) = 2$

Step 6:   Because the logarithm of 36 can be taken, -6 is a valid solution. Now the value 5 can be plugged in as follows.

$\log_6 \left(x^2 + x + 6\right) = 2$

$\log_6 \left(5^2 + 5 + 6\right) = 2$

$\log_6 (25 + 5 + 6) = 2$

$\log_6 (36) = 2$

Again, because the logarithm of 36 can be taken, 5 is also a valid solution. Therefore, the solution to the equation is $x = -6$ and 5.

Example 18: Solve the equation $\log_4 (x - 6) + \log_4 (x) = 2$ for x.

Step 1: Use the property $\log_a (cd) = \log_a c + \log_a d$.

The property $\log_a (cd) = \log_a c + \log_a d$, can be used to simplify the equation $\log_4 (x - 6) + \log_4 (x) = 2$ as follows.

$$\log_4 (x - 6) + \log_4 (x) = 2 \rightarrow \log_4 ((x - 6) \cdot x) = 2 \rightarrow \log_4 (x^2 - 6x) = 2$$

Step 2: Use the fact that another way of writing $\log_a x = y$ is $x = a^y$.

Since another way of writing $\log_a x = y$ is $x = a^y$, the equation can also be written as $x^2 - 6x = 4^2$, or $x^2 - 6x = 16$.

Step 3: Write the equation $x^2 - 6x = 16$ in standard form.

$$x^2 - 6x - 16 = 0$$

Step 4: Factor the left side of the equation.

$$(x + 2)(x - 8) = 0$$

Step 5: Solve for x.

Since $(x + 2)(x - 8) = 0$, either $x + 2 = 0$ or $x - 8 = 0$. This mean that x can be solved for as follows.

$$x + 2 = 0 \qquad x - 8 = 0$$
$$x = -2 \qquad x = 8$$

Step 6: Check the values of x to make sure that only the logarithm of positive numbers are being taken.

Since only the logarithm of positive numbers can be taken, the values of x must be plugged into the original equation to make sure that they are valid solutions. The value -2 can be plugged in as follows.

$$\log_4 (x - 6) + \log_4 (x) = 2$$

$$\log_4 (-2 - 6) + \log_4 (-2) = 2$$

$$\log_4 (-8) + \log_4 (-2) = 2$$

Step 7: Because the logarithm of -8 and -2 cannot be taken, -2 is not a valid solution. Now the value 8 can be plugged in as follows.

$$\log_4 (x - 6) + \log_4 (x) = 2$$

$$\log_4 (8 - 6) + \log_4 (8) = 2$$

$$\log_4 (2) + \log_4 (8) = 2$$

Because the logarithm of 2 and 8 can be taken, 8 is a valid solution. Therefore, the solution to the equation is $x = 8$.

Example 19:   Solve the equation $\ln(x+5) + \ln(x+2) = 2\ln(x+3)$ for x.

Step 1:   Use the property $\log_a(cd) = \log_a c + \log_a d$.

The property $\log_a(cd) = \log_a c + \log_a d$, can be used to simplify the original equation as follows.

$$\ln(x+5) + \ln(x+2) = 2\ln(x+3) \rightarrow \ln((x+5)(x+2)) = 2\ln(x+3)$$
$$\rightarrow \ln(x^2 + 7x + 10) = 2\ln(x+3)$$

Step 2:   Use the property $\log_a(c^d) = d\log_a c$ to further simplify the equation.

$$\ln(x^2 + 7x + 10) = 2\ln(x+3)$$
$$\ln(x^2 + 7x + 10) = \ln(x+3)^2$$
$$\ln(x^2 + 7x + 10) = \ln((x+3)(x+3))$$
$$\ln(x^2 + 7x + 10) = \ln(x^2 + 6x + 9)$$

Step 3:   Use the property that says if $\log_a c = \log_a d$, then $c = d$.

Since $\ln(x^2 + 7x + 10) = \ln(x^2 + 6x + 9)$, it must be true that $x^2 + 7x + 10 = x^2 + 6x + 9$.

Step 4:   Solve for x.

$$x^2 + 7x + 10 = x^2 + 6x + 9 \rightarrow 7x + 10 = 6x + 9 \rightarrow x = -1$$

Step 5:   Check the value of x.

$$\ln(x+5) + \ln(x+2) = 2\ln(x+3) \rightarrow \ln(-1+5) + \ln(-1+2) =$$
$$2\ln(-1+3) \rightarrow \ln(4) + \ln(1) = 2\ln(2)$$

Because logarithm of 4, 1, and 2 can be taken, -1 is a valid solution. Therefore, the solution to the equation $\ln(x+5) + \ln(x+2) = 2\ln(x+3)$ is $x = -1$.

Solve each of the following equations for x.

1. $\log_5(x^2 - 2x - 3) = 1$

2. $\log_2(x^2 + 10x + 53) = 5$

3. $\log(x^2 - 8x + 91) = 2$

4. $\log_3(x^2 - x + 25) = 4$

5. $\log_4(x^2 - 9x + 82) = 3$

6. $\log_7(x^2 + 3x - 21) = 1$

7. $\log_2(x - 30) + \log_2(x) = 6$

8. $\log_6(x - 5) + \log_6(x) = 2$

9. $\log_8(x^3 + 2x^2) - \log_8(x) = 1$

10. $\log(x^3 - 21x^2) - \log(x) = 2$

11. $2\log_4(x + 6) = 1$

12. $2\log_5(x - 1) = 2$

13. $\ln(x + 3) + \ln(x + 2) = \ln(x^2 + 4x + 10)$

14. $2\log(x - 2) = \log(x^2 - 20)$

15. $\log_2(2x^2 - 10x + 12) = \log_2(x^2 + 3x - 28)$

16. $\log_5(x + 1) + \log_5(x + 12) = 2\log_5(x + 6)$

17. $\ln(x^3 + 3x^2 + 2x + 6) = 3\ln(x + 1)$

18. $2\log_9(x + 4) = \log_9(x + 2) + \log_9(x + 7)$

6.8   Solving Logarithmic Equations Graphically

There are two ways to solve a logarithmic equation graphically.

1) Graph the left side of the equation, then graph the right side, and determine the x-coordinates of the points where the graphs intersect. If there is no intersection, the equation has no solution.

2) Move all the terms of the logarithmic equation to the left side, graph the left side, and determine the x-coordinates of the points where the graph crosses the x-axis. If the graph does not cross the x-axis, then the equation has no solution.

Remember to use the logarithmic property $\log_a c = \dfrac{\log c}{\log a}$ when graphing.

Example 20:   Solve the equation $\log_2 (x^2 + 5x + 6) = 1$ graphically.

Step 1:   Use the property $\log_a c = \dfrac{\log c}{\log a}$.

The equation $\log_2 (x^2 + 5x + 6) = 1$ can be written as $\dfrac{\log (x^2 + 5x + 6)}{\log (2)} = 1$.

Step 2:   Graph the left side of the equation.

First, the function $f(x) = \dfrac{\log (x^2 + 5x + 6)}{\log (2)}$ should be graphed as follows.

Calculator:

a. Press ⟨ Y = ⟩.

b. Enter log(X^2 + 5X + 6)/log(2) after $Y_1 =$.

c. Press ⟨ WINDOW ⟩.

d. Enter the following values.
Xmin = –5
Xmax = 5
Xscl = 1
Ymin = –5
Ymax = 5
Yscl = 1
Xres = 1

e. Press ⟨ GRAPH ⟩.

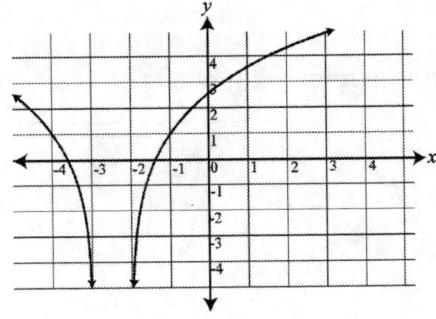

Step 3:   Graph the right side of the equation on the same coordinate grid.

Next, the function $g(x) = 1$ should be graphed on the same coordinate grid as follows.

> **Calculator:**
> a. Press $\boxed{Y =}$.
>
> b. Enter log(X^2 + 5X + 6)/log(2) after $Y_1 =$ and 1 after Y_2.
>
> c. Press $\boxed{\text{WINDOW}}$.
>
> d. Enter the values from Step 2.
>
> e. Press $\boxed{\text{GRAPH}}$.

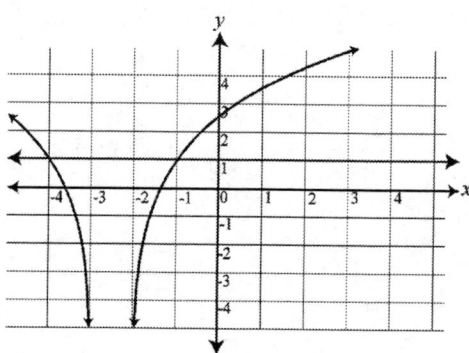

Step 4:   Determine the x-coordinates of the points where the graphs intersect.

> **Calculator:**
> a. Press $\boxed{\text{2ND}}$ $\boxed{\text{TRACE}}$.
>
> b. Press $\boxed{5}$.
>
> c. Press $\boxed{\text{ENTER}}$ twice.
>
> d. Move the cursor as close as possible to a point of intersection.
>
> e. Press $\boxed{\text{ENTER}}$.
>
> f. Repeat the steps above for the remaining points of intersection.

Since the graphs intersect at the points $(-4, 1)$ and $(-1, 1)$, the solution to the equation $\log_2(x^2 + 5x + 6) = 1$ is $x = -4$ and -1.

Example 21:   Solve the equation $\log(x+4) + \log(x-1) = 2\log(x+1)$ graphically.

Step 1:   Move all the terms to the left side of the equation.

When all the terms of $\log(x+4) + \log(x-1) = 2\log(x+1)$ are moved to the left side of the equation, the right side of the equation becomes 0 as follows.

$$\log(x+4) + \log(x-1) - 2\log(x+1) = 0$$

Step 2:   Graph the left side of the equation.

Now, the function $f(x) = \log(x+4) + \log(x-1) - 2\log(x+1)$ should be graphed as follows.

Calculator:

a. Press $\boxed{Y=}$.

b. Enter $\log(X+4) + \log(X-1) - 2 * \log(X+1)$ after $Y_1 = $.

c. Press $\boxed{WINDOW}$.

d. Enter the following values.

Xmin = –2
Xmax = 8
Xscl = 1
Ymin = –0.07
Ymax = 0.03
Yscl = 0.01
Xres = 1

e. Press $\boxed{GRAPH}$.

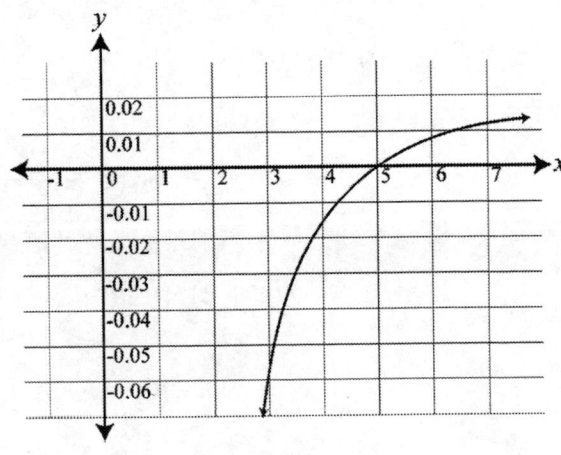

Step 3:   Determine the x-coordinate of the point where the graph crosses the x-axis.

Calculator:

a. Press [2ND] [TRACE].

b. Press [2].

c. Move the cursor to the left of the point where the graph crosses the x-axis.

d. Press [ENTER].

e. Move the cursor to the right of the point where the graph crosses the x-axis.

f. Press [ENTER] twice.

Since the graph crosses the x-axis at the point $(5, 0)$, the solution to the equation $\log(x + 4) + \log(x - 1) - 2\log(x + 1)$ is $x = 5$.

Solve each of the following equations graphically.

1. $\log_4(x^2 + 7x + 12) = \frac{1}{2}$

2. $\log_3(x^2 + 12x + 20) = 2$

3. $\log_2(x^2 + 2x - 8) = 4$

4. $\log_6(x^2 - 6x + 9) = 2$

5. $\log_{12}(x^2 - 3x + 2) = 1$

6. $\log_2(x^2 - 4x + 3) = 3$

7. $\log(x + 6) + \log(x + 3) = 2\log(x + 4)$

8. $\log(x - 4) + \log(x - 1) = 2\log(x - 3)$

9. $\ln(x + 10) + \ln(x + 1) = 2\ln(x + 5)$

10. $2\ln(x + 6) = \ln(x + 9) + \ln(x + 4)$

11. $\log(x - 5) + \log(x + 5) = 2\log(x - 1)$

12. $2\log(x + 3) = \log(x + 1) + \log(x + 6)$

Example 22: Determine if the equation $\log_9\left(x^2 + 27\right) = 1$ has a solution.

Step 1: Use the property $\log_a c = \dfrac{\log c}{\log a}$.

The equation $\log_9\left(x^2 + 27\right) = 1$ can be rewritten as $\dfrac{\log\left(x^2 + 27\right)}{\log\left(9\right)} = 1$.

Step 2: Move all the terms to the left side of the equation.

When all the terms of $\log_9\left(x^2 + 27\right) = 1$ are moved to the left side of the equation, the right side of the equation becomes 0 as follows.

$$\dfrac{\log\left(x^2 + 27\right)}{\log\left(9\right)} - 1 = 0$$

Step 3: Graph the left side of the equation.

Now, the function $f\left(x\right) = \dfrac{\log\left(x^2 + 27\right)}{\log\left(9\right)} - 1$ should be graphed as follows.

Calculator:
a. Press $\boxed{Y =}$.

b. Enter log(X^2 + 27)/log(9) − 1 after $Y_1 =$.

c. Press $\boxed{\text{GRAPH}}$.

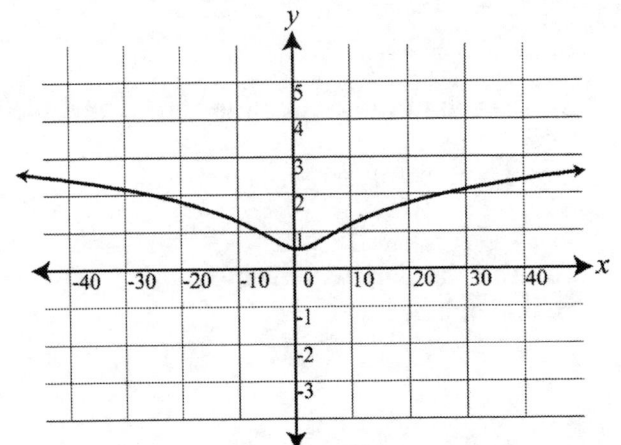

Step 4: Determine if the graph crosses the x-axis.

Since the graph doesn't cross the x-axis, it doesn't have a solution.

Determine if each of the following equations has a solution.

1. $\log_5\left(x^2 + 10\right) = 2$ 3. $\log_3\left(x^2 + 77\right) = 4$ 5. $\log\left(x^2 + 91\right) = 2$

2. $\log_4\left(x^2 + 15\right) = 3$ 4. $\log_{16}\left(x^2 + 58\right) = 1$ 6. $\log_2\left(x^2 + 64\right) = 5$

Chapter 6 Review

Find real and complex roots of higher degree polynomial equations.

1. $x^3 - 2x^2 - 13x - 10 = 0$

2. $x^3 + 6x^2 + 5x - 12 = 0$

3. $x^3 + 4x^2 - 5x - 20 = 0$

4. $x^3 - 3x^2 - 6x + 18 = 0$

Find all of the roots of each of the following equations. For each equation, one of the roots has been given.

5. $x^3 + 2x^2 + 36x + 72 = 0$; $x = 6i$

6. $x^4 - 9x^3 + 24x^2 - 36x + 80 = 0$; $x = -2i$

Solve each of these equations for x.

7. $x^2 + 8x = 48$

8. $x^2 + 14x = -45$

9. $x^3 + 6x^2 = 19x + 84$

10. $x^3 = 4x^2 + 29x + 24$

Solve each of these equations for x. Include both real and complex solutions.

11. $x^4 + 6x^3 + 13x^2 + 24x = -36$

12. $x^4 + 56x = 4x^3 + 10x^2 + 56$

To solve the polynomial equation $x^3 - 6x^2 = x - 30$, both sides of the equation were graphed as shown below.

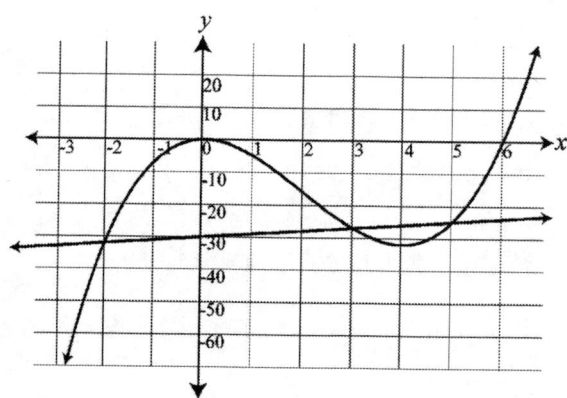

13. What two functions were graphed?

14. What is the solution to the polynomial equation $x^3 - 6x^2 = x - 30$?

To solve the polynomial equation $x^4 + x + 12 = x^3 + 13x^2$, the equation was written in standard form and then the left side of the equation was graphed as shown below.

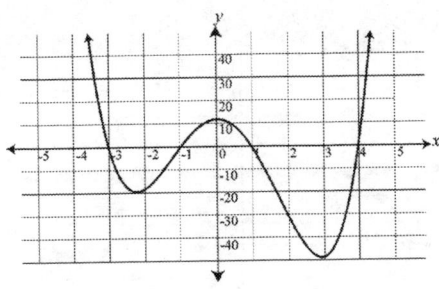

15. What function was graphed?

16. What is the solution to the polynomial equation $x^4 + x + 12 = x^3 + 13x^2$?

Determine the number of distinct real solutions to each of the following equations.

17. $x^3 + 25x = 3x^2 + 75$

18. $x^6 + x^4 + 49x^2 = -49$

Solve each of the following equations for x. Round your answers to two decimal places.

19. $4^x = 112$

20. $\left(\frac{3}{5}\right)^x = 77$

21. $\left(\frac{1}{5}\right)\left(9^{4x-1}\right) - 33 = 102$

22. $(8)\left(2^{x+11}\right) + 19 = 75$

23. $7e^{x-6} + 36 = 85$

24. $2e^{\frac{x+32}{2}} - 57 = 213$

To solve the exponential equation $(2)\left(4^{x+3}\right) - 23 = 9$, both sides of the equation were graphed as shown below.

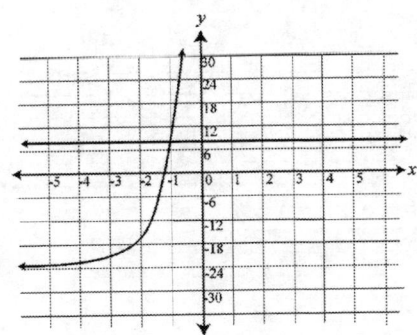

25. What two functions were graphed?

26. What is the solution to the exponential equation $(2)\left(4^{x+3}\right) - 23 = 9$?

To solve the exponential equation $\frac{3}{4}\left(6^{3x-1}\right) + 15 = 42$, all the terms were moved to the left side of the equation and then the left side of the equation was graphed as shown below.

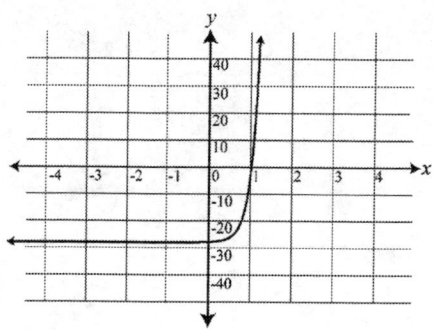

27. What function was graphed?

28. What is the solution to the exponential equation $\frac{3}{4}\left(6^{3x-1}\right) + 15 = 42$?

Determine if each of the following equations has a solution.

29. $(2)\left(7^{x-5}\right) + 106 = 8$

30. $(75)\left(5^{x+3}\right) - 20 = -17$

Solve each of the following equations for x.

31. $\log_3\left(x^2 + 4x - 5\right) = 3$

32. $\log_9\left(x^2 + 2x - 15\right) = 1$

33. $\log_8\left(x + 12\right) + \log_8\left(x\right) = 2$

34. $2\log_6\left(x - 10\right) = 2$

35. $\ln\left(x + 4\right) + \ln\left(x + 5\right) = \ln\left(x^2 + 8x + 25\right)$

36. $\log\left(2x^2 - 13x + 62\right) = \log\left(x^2 + 2x + 8\right)$

To solve the logarithmic equation $\log_2\left(x^2 + 10x + 25\right) = 4$, both sides of the equation were graphed as shown below.

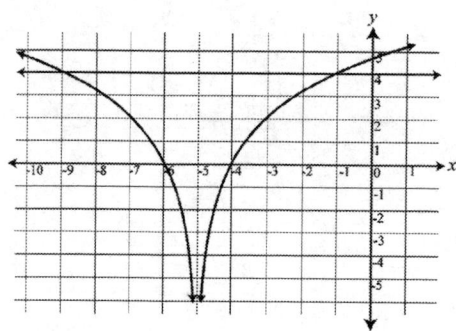

37. What two functions were graphed?

38. What is the solution to the logarithmic equation $\log_2\left(x^2 + 10x + 25\right) = 4$?

To solve the logarithmic equation $\log(x-6) + \log(x-1) = 2\log(x-4)$, **all the terms were moved to the left side of the equation and then the left side of the equation was graphed as shown below.**

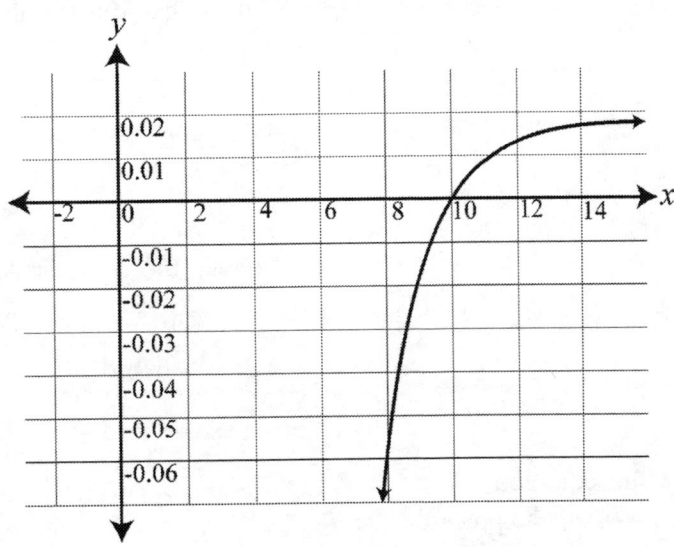

39. What function was graphed?

40. What is the solution to the logarithmic equation $\log(x-6) + \log(x-1) = 2\log(x-4)$?

Determine if each of the following equations has a solution.

41. $\log_9(x^2 + 66) = 2$

42. $\log_{22}(x^2 + 35) = 1$

Chapter 6 Test

1. According to the Rational Root Theorem, which of these is a possible rational root of the equation $2x^3 + 3x^2 - 23x - 12 = 0$?

   A. $x = \dfrac{1}{12}$

   B. $x = \dfrac{1}{6}$

   C. $x = \dfrac{1}{4}$

   D. $x = \dfrac{1}{2}$

2. To find the roots of the equation $x^3 - 8x^2 - 3x + 90 = 0$, all the possible rational roots were tested with the function $f(x) = x^3 - 8x^2 - 3x + 90$. Is $x = -3$ a root of the equation?

   A. No, because $f(-3) \neq 0$.
   B. No, because $f(-3) = 0$.
   C. Yes, because $f(-3) \neq 0$.
   D. Yes, because $f(-3) = 0$.

3. What are the roots of the equation $x^3 - 19x + 30 = 0$?

   A. $x = -5, -2,$ and 3
   B. $x = -5, 2,$ and 3
   C. $x = -2, 3,$ and 5
   D. $x = 2, 3,$ and 5

4. What is the conjugate of $5i$?

   A. $-5i$

   B. $-\dfrac{i}{5}$

   C. $\dfrac{i}{5}$

   D. $5i$

5. One of the solutions of an equation is $x = 6$, and this solution occurs twice. What is the multiplicity of this solution?

   A. 2
   B. 3
   C. 6
   D. 12

6. What is the solution to the equation $3x^2 = 17x + 6$?

   A. $x = -6$ and $-\frac{1}{3}$

   B. $x = -6$ and $\frac{1}{3}$

   C. $x = -\frac{1}{3}$ and 6

   D. $x = \frac{1}{3}$ and 6

7. If $x^3 + 6x^2 - 19x - 24$ is divided by $x + 1$, what is the result?

   A. $x^2 - 5x - 24$
   B. $x^2 - 5x + 24$
   C. $x^2 + 5x - 24$
   D. $x^2 + 5x + 24$

8. What is the solution to the equation $x^3 + 12x^2 - 2x - 24 = 0$?

   A. $x = -12, -2i,$ and $2i$
   B. $x = -12, -\sqrt{2},$ and $\sqrt{2}$
   C. $x = 12, -2i,$ and $2i$
   D. $x = 12, -\sqrt{2},$ and $\sqrt{2}$

9. The graphs of the functions $f(x) = x^2$ and $g(x) = 3x + 28$ intersect at the points $(-4, 16)$ and $(7, 49)$. What is the solution to the equation $x^2 = 3x + 28$?

   A. $x = -4$ and 16
   B. $x = -4$ and 7
   C. $x = 7$ and 49
   D. $x = 16$ and 49

10. To solve the equation $x^3 - 2x^2 - 29x - 42 = 0$, the function $f(x) = x^3 - 2x^2 - 29x - 42$ was graphed. At what points does the graph of the function cross the x-axis?

    A. $(-3, 0)$, $(-2, 0)$, and $(-7, 0)$
    B. $(-3, 0)$, $(-2, 0)$, and $(7, 0)$
    C. $(0, -3)$, $(0, -2)$, and $(0, -7)$
    D. $(0, -3)$, $(0, -2)$, and $(0, 7)$

11. The graph of the function $f(x) = x^3 + 5x^2 + 2x - 8$ is shown below.

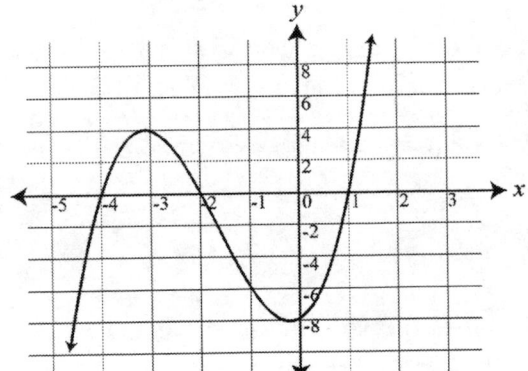

    What is the solution to the equation $x^3 + 5x^2 + 2x - 8 = 0$?

    A. $x = -4, -2,$ and 1
    B. $x = -4, 2,$ and 1
    C. $x = -2, -1,$ and 4
    D. $x = -2, 1,$ and 4

12. How many distinct real solutions does the equation $x^5 - x^4 + 10x^3 - 10x^2 + 9x - 9 = 0$ have?

    A. 1
    B. 3
    C. 4
    D. 5

13. What is the solution to the equation $\left(\frac{1}{4}\right)^x = 150$ rounded to two decimal places?

    A. $x = -3.61$
    B. $x = -0.28$
    C. $x = 0.28$
    D. $x = 3.61$

14. When solving the equation $15^x = 905$, Jenny's first step was to write $\log(15^x) = \log(905)$, Mike's was to write $\ln(15^x) = \ln(905)$, and Bob's was to write $\log_{15}(15^x) = \log_{15}(905)$. Whose equation was correct?

    A. Only Mike's
    B. Only Jenny's and Mike's
    C. Only Bob's and Mike's
    D. Jenny's, Bob's, and Mike's

15. What is the solution to the equation $(7)(2^{x-4}) + 25 = 95$ rounded to two decimal places?

    A. $x = 0.30$
    B. $x = 3.32$
    C. $x = 4.30$
    D. $x = 7.32$

16. If $e^{x+1} = 99$, which of these expressions is equal to the value of x?

    A. $\ln(98)$
    B. $\ln(99) - 1$
    C. $\ln(99) + 1$
    D. $\ln(100)$

17. The equation $(2)(5^x) - 79 = 171$ was solved by moving all the terms to the left side of the equation and then graphing the left side. Which of these functions was graphed?

    A. $f(x) = (2)(5^x) - 250$
    B. $f(x) = (2)(5^x) - 92$
    C. $f(x) = (2)(5^x) + 92$
    D. $f(x) = (2)(5^x) + 250$

18. The graphs of the functions $f(x) = (4)(9^x) - 26$ and $g(x) = 10$ are shown below.

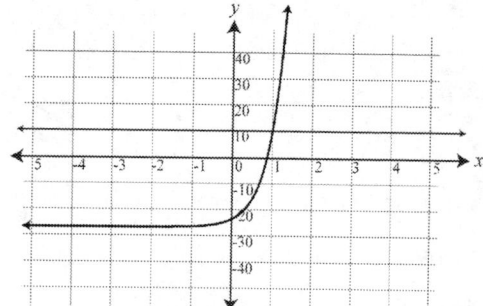

What is the solution to the equation $(4)(9^x) - 26 = 10$?

A. $x = -10$
B. $x = -1$
C. $x = 1$
D. $x = 10$

19. How many times does the graph of the equation $\left(\frac{1}{3}\right)(21^x) + 8 = 1$ intersect the x-axis?

A. 0
B. 1
C. 2
D. 3

20. The graph of the function $f(x) = (24)(4^x) - 6$ is shown below.

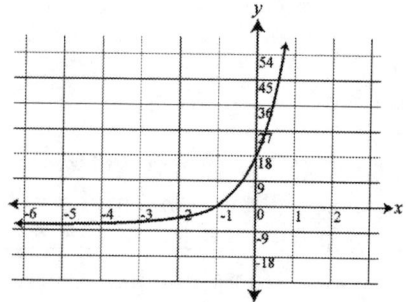

What is the solution to the equation $(24)(4^x) - 6 = 0$?

A. $x = -18$
B. $x = -1$
C. $x = 1$
D. $x = 18$

21. Which of these is a correct statement?

A. It's not possible to take the logarithm of neither positive numbers nor negative numbers.

B. It's possible to take the logarithm of only positive numbers.

C. It's possible to take the logarithm of only negative numbers.

D. It's possible to take the logarithm of both positive numbers and negative numbers.

22. For how many values of x is the equation $\log_2(x - 7) + \log_2(x) = 3$ valid?

A. 0
B. 1
C. 2
D. 3

23. What is the solution to the equation $\log(x^3 + 3x^2) - \log(x) = 1$?

A. $x = -5$
B. $x = -2$
C. $x = 2$
D. $x = 5$

24. If $\log_a c = \log_a d$, then c must be equal to what?

A. 0
B. 1
C. a
D. d

25. Which of these expressions is equal to $\log_a c$?

A. $\dfrac{\log a}{\log c}$

B. $\dfrac{\log c}{\log a}$

C. $\log a - \log c$

D. $\log c = \log a$

Chapter 7
Solving Polynomial, Exponential, and Logarithmic Inequalities

This chapter covers the following Georgia Performance Standards:

MM3A	Algebra	MM3A3c, MM3A3d

7.1 Solving Polynomial Inequalities Analytically

To solve a polynomial inequality analytically, first replace the inequality sign with an equal sign to produce a polynomial equation. Then, write the polynomial equation in standard form, which means that the terms on the left side of the equation are in order from the highest degree to the lowest degree and the right side of the equation is 0. The real roots of the equation can then be found by simply factoring, or if necessary, by using tools such as the Rational Root Theorem, the Factor Theorem, polynomial division, and the quadratic formula. Once the real roots are found, different regions of a number line can be tested to determine which regions satisfy the original inequality, and the solution to the inequality can be written in interval notation.

Example 1:   Write the solution to the inequality $x^2 - x < 6$ in interval notation.

   Step 1:   Replace the inequality sign with an equals sign to produce a polynomial equation.

   When $<$ is replaced with an equal sign, the inequality $x^2 - x < 6$ becomes $x^2 - x = 6$.

   Step 2:   Write the polynomial equation in standard form.

   To write the equation $x^2 - x = 6$ in standard form, the terms on the left side of the equation should be in order from the highest degree to the lowest degree and the right side of the equation should be 0. The equation can be changed to standard form as follows.

   $x^2 - x = 6$

   $x^2 - x - 6 = 6 - 6$

   $x^2 - x - 6 = 0$

Step 3:   Factor the left side of the equation.

Since the equation is quadratic, with the middle term on the left side having a coefficient of -1 and the last term being -6, the left side can be factored as follows.

$$x^2 - x - 6 = 0$$

$$(x + 2)(x - 3) = 0$$

Step 4:   Find the real roots of the equation.

Since $(x + 2)(x - 3) = 0$, either $x + 2 = 0$ or $x - 3 = 0$. This means that the real roots of the equation can be found as follows.

$$x + 2 = 0$$

$$x + 2 - 2 = 0 - 2$$

$$x = -2$$

and

$$x - 3 = 0$$

$$x - 3 + 3 = 0 + 3$$

$$x = 3$$

Therefore, the real roots of the equation are $x = -2$ and 3.

Step 5:   Divide a number line into different regions.

Now that the real roots of the equation are known, they can be plotted on a number line, and the number line can be divided into three different regions as follows.

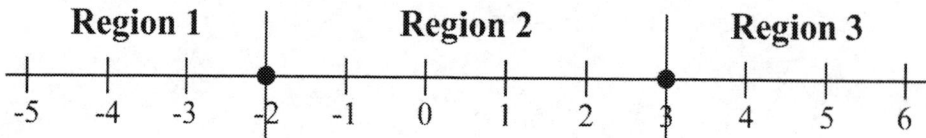

Copyright © American Book Company

Step 6:   Test each of the regions.

Any number from each of the regions can be tested with the original inequality to determine which regions satisfy the inequality. First, the number -4 from region 1 can be tested as follows.

$x^2 - x < 6$

$(-4)^2 - (-4) < 6$

$16 + 4 < 6$

$20 < 6$

Since the inequality $20 < 6$ is false, region 1 does not satisfy the inequality.

Step 7:   Next, the number 1 from region 2 can be tested as follows.

$x^2 - x < 6$

$1^2 - 1 < 6$

$1 - 1 < 6$

$0 < 6$

Since the inequality $0 < 6$ is true, region 2 satisfies the inequality.

Step 8:   Finally, the number 5 from region 3 can be tested as follows.

$x^2 - x < 6$

$(5)^2 - 5 < 6$

$25 - 5 < 6$

$20 < 6$

Since the inequality $20 < 6$ is false, region 3 does not satisfy the inequality.

Step 9:   Write the solution to the inequality in interval notation.

Because the only region that satisfies the inequality is region 2, x must be greater than -2, but less than 3. Note that x is not equal to -2 or 3, since there is $<$ in the original inequality and not $\leq$. Therefore, the solution to the inequality in interval notation is $(-2, 3)$.

Write the solution to each of the following inequalities in interval notation.

1. $x^2 - 2x < 24$

2. $x^2 + 5 > -6x$

3. $x^2 + 3x \leq 70$

4. $18 + x^2 \geq 11x$

5. $x^2 < 5x + 24$

6. $4x^2 > 15x + 4$

7. $x^2 \leq x + 42$

8. $x^2 + 64 > 16x$

9. $9x^2 + 3x \geq 2$

Example 2:   Write the solution to the inequality $x^3 + x^2 \geq 10x + 10$ in interval notation.

Step 1:   Replace the inequality sign with an equal sign to produce a polynomial equation.

When $\geq$ is replaced with $=$, the inequality $x^3 + x^2 \geq 10x + 10$ becomes $x^3 + x^2 = 10x + 10$.

Step 2:   Write the polynomial equation in standard form.

To write the equation $x^3 + x^2 = 10x + 10$ in standard form, the terms on the left side of the equation should be in order from the highest degree to the lowest degree and the right side of the equation should be 0. The equation can be changed to standard form as follows.

$x^3 + x^2 = 10x + 10$

$x^3 + x^2 - 10x - 10 = 10x + 10 - 10x - 10$

$x^3 + x^2 - 10x - 10 = 0$

Step 3:   Use the Rational Root Theorem to find all of the possible rational roots of the equation.

Since the equation is now in standard form, all of its possible rational roots can be found by dividing all of the positive and negative factors of the last term on the left side by all of the positive and negative factors of the coefficient of the first term. Since the last term on the left side is 10, all of its positive and negative factors are ± 1, ± 2, ± 5, and ± 10. Also, since the coefficient of the first term is 1, all of its positive and negative factors are ± 1. This means that all of the possible rational roots of the equation are

$x = \dfrac{\pm 1}{\pm 1}, \dfrac{\pm 2}{\pm 1}, \dfrac{\pm 5}{\pm 1}$, and $\dfrac{\pm 10}{\pm 1}$, or $x = \pm 1$, ± 2, ± 5, and ± 10.

Step 4:   Use the Factor Theorem to find a rational root.

According to the Factor Theorem, if a root of the equation is plugged in for x, the left side of the equation becomes 0. With this in mind, all of the possible rational roots of the equation can be tested. First, $x = 1$ and $x = -1$ can be tested as follows.

$f(x) = x^3 + x^2 - 10x - 10$

$f(x) = (1)^3 + (1)^2 - 10(1) - 10 = 1 + 1 - 10 - 10 = -18$

$f(-1) = (-1)^3 + (-1)^2 - 10(-1) - 10 = -1 + 1 + 10 - 1 = 00$

This means that $x = -1$ is a root of the equation, but $x = 1$ is not. At this point, only one of the roots of the equation has been found, but the Fundamental Theorem of Algebra states that the equation has 3 roots, since the highest exponent on the left side of the equation is 3. While it's possible to keep using the Factor Theorem to test the other possible rational roots, a faster approach may be to produce a quadratic equation to find the two additional roots.

Step 5:   Use polynomial division to produce a quadratic equation.

Since $x = -1$ is a root of the equation $x^3 + x^2 - 10x - 10 = 0$, a factor of $x^3 + x^2 - 10x - 10$ must be $x + 1$, so $x^3 + x^2 - 10x - 10$ can be divided by $x + 1$ to produce one side of the quadratic equation. The division is done synthetically as follows.

$$
\begin{array}{r|rrrr}
-1 & 1 & 1 & -10 & -10 \\
   &   & -1 & 0 & 10 \\
\hline
   & 1 & 0 & -10 & 0
\end{array}
$$

Since $x^3 + x^2 - 10x - 10$ divided by $x + 1$ equals $x^2 - 10$, the quadratic equation that is produced is $x^2 - 10 = 0$.

Step 6:   Use the quadratic formula to find the remaining real roots.

Now that there is a quadratic equation in the form $ax^2 + bx + c = 0$ the quadratic formula can be used to solve it. The quadratic formula is $x = \dfrac{-b \pm \sqrt{b^2 - 4ac}}{2a}$, and in this case, $a = 1$, $b = 0$, and $c = -10$. Thus, remaining real roots can be found as follows.

$$x = \frac{-b \pm \sqrt{b^2 - 4ac}}{2a} = \frac{0 \pm \sqrt{0^2 - 4\,(1)\,(-10)}}{2\,(1)} = \frac{\pm\sqrt{40}}{2} = \frac{\pm 2\sqrt{10}}{2} = \pm\sqrt{10}$$

This means that three real roots of the equation $x^3 + x^2 - 10x - 10 = 0$ are $x = -1, -\sqrt{10},$ and $\sqrt{10}$.

Step 7:   Divide a number line into different regions.

Now that the real roots of the equation are known, they can be plotted on a number line, and the number line can be divided into four different regions as follows.

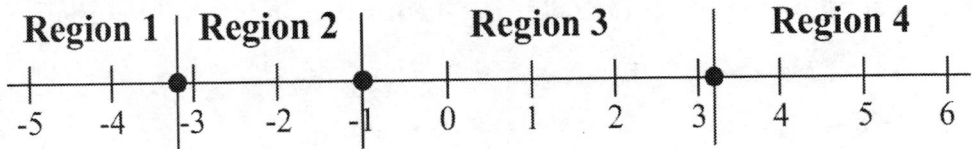

Step 8:   Test each of the regions.

Any number from each of the regions can be tested with the original inequality to determine which regions satisfy the inequality. First, the number -5 from region 1 can be tested as follows.

$(-5)^3 + (-5)^2 \geq 10(-5) + 10$

$-125 + 25 \geq -50 + 10$

Since the inequality $-100 \geq -40$ is false, region 1 does not satisfy the inequality.

Step 9:   Next, the number -2 from region 2 can be tested as follows.

$(-2)^3 + (-2)^2 \geq 10(-2) + 10$

$-8 + 4 \geq -20 + 10$

$-4 \geq -10$

Since the inequality $-4 \geq -10$ is true, region 2 satisfies the inequality.

Step 10:   Next, the number 1 from region 3 can be tested as follows.

$(1)^3 + (1)^2 \geq 10(1) + 10$

$1 + 1 \geq 10 + 10$

$2 \geq 20$

Since the inequality $2 \geq 20$ is false, region 3 does not satisfy the inequality.

Step 11:   Finally, the number 5 from region 4 can be tested as follows.

$(5)^3 + (5)^2 \geq 10(5) + 10$

$125 + 25 \geq 50 + 10$

$150 \geq 60$

Since the inequality $150 \geq 60$ is false, region 4 satisfies the inequality.

Step 12:   Write the solution to the inequality in interval notation.

Because the only regions that satisfy the inequality are region 2 and region 4, x must be greater than or equal to $-\sqrt{10}$ and less than or equal to -1, or it must be greater than or equal to $\sqrt{10}$. Therefore, the solution to the inequality in interval notation is $\left[-\sqrt{10}, -1\right] \cup \left[\sqrt{10}, \infty\right)$.

Write the solution to each of the following inequalities in interval notation.

1. $x^3 + 9x^2 + 23x < -15$

2. $14x + x^3 + 24 > 9x^2$

3. $x^3 + 5x^2 \leq 4x + 20$

4. $x^3 + 7x^2 \geq 63 + 9x$

5. $32 + x^3 < 2x^2 + 16x$

6. $2x^2 + x^3 > 30 + 15x$

7. $x^3 + 6x^2 \leq 8x + 48$

8. $25x + x^3 \geq 2x^2 + 50$

9. $x^3 + 16x + 3x^2 < -48$

7.2   Solving Polynomial Inequalities Graphically

There are two ways to solve a polynomial inequality graphically.

1) Graph the left side of the inequality, graph the right side of the inequality on the same coordinate grid, and determine the regions of the graph that satisfy the inequality.

2) Write the polynomial inequality in standard form, graph the left side of the inequality, and determine the regions of the graph that satisfy the inequality.

With either approach, if no region of the graph satisfies the inequality, then the inequality has no solution.

Example 3:   Solve the inequality $x^3 - 3x^2 \leq 13x - 15$ graphically. Write the solution in interval notation.

Step 1:   Write the inequality in standard form.

To write the inequality $x^3 - 3x^2 \leq 13x - 15$ in standard form, the terms on the left side of the inequality should be in order from the highest degree to the lowest degree and the right side of the inequality should be 0. The inequality can be converted to standard form as follows.

$$x^3 - 3x^2 - 13x + 15 \leq 13x - 15 - 13x + 15$$
$$x^3 - 3x^2 - 13x + 15 \leq 0$$

Step 2:   Graph the left side of the inequality.

Now, the function $f(x) = x^3 - 3x^2 - 13x + 15$ should be graphed as followed.

Calculator:
On the TI-83/84 graphing calculator, this graph can be produced by performing the following steps.

a. Press $\boxed{Y =}$.

b. Enter X^3 – 3X^2 – 13X + 15 after $Y_1 =$.

c. Press $\boxed{\text{WINDOW}}$.

d. Enter the following values.

Xmin = –6
Xmax = 6
Xscl = 1
Ymin = –30
Ymax = 30
Yscl = 5
Xres = 1

e. Press $\boxed{\text{GRAPH}}$.

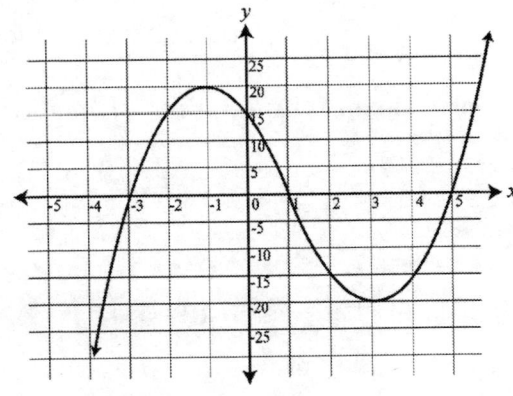

Step 3:   Determine the x-coordinates of the points where the graph crosses the x-axis.

> **Calculator:**
>
> On the TI-83/84 graphing calculator, the points where the graph crosses the x-axis can be found by performing the following steps.
>
> a. Press $\boxed{\text{2ND}}$ $\boxed{\text{TRACE}}$.
>
> b. Press $\boxed{2}$.
>
> c. Move the cursor to the left of a point where the graph crosses the x-axis.
>
> d. Press $\boxed{\text{ENTER}}$.
>
> e. Move the cursor to the right of a point where the graph crosses the x-axis.
>
> f. Press $\boxed{\text{ENTER}}$ twice.
>
> g. Repeat the steps above for the remaining points of intersection.

The graph crosses the x-axis at the points $(-3, 0)$, $(1, 0)$, and $(5, 0)$, so the x-coordinates are -3, 1, and 5.

Step 4:   Determine the regions of the graph that satisfy the inequality.

Since the inequality in standard form $x^3 - 3x^2 - 13x + 15 \leq 0$, the regions of the graph that satisfy the inequality are those in which the graph of $f(x) = x^3 - 3x^2 - 13x + 15$ is at or below the x-axis. This occurs when x is less than or equal to -3, and also when x is greater than or equal to 1 and less than or equal to 5.

Step 5:   Write the solution to the inequality in interval notation.

Therefore, the solution to the inequality in interval notation is $(\infty, -3] \cup [1, 5]$.

Solve each of the following inequalities graphically. Write each solution in interval notation.

1. $x^3 + x^2 > 37x - 35$

2. $x^3 + 3x^2 < 40x + 84$

3. $x^3 - 11x \geq 2x^2 - 12$

4. $x^3 + 90 \leq 6x^2 + 37x$

5. $x^4 + 2x^3 - 35x^2 > 72x + 36$

6. $x^4 + 2x^3 - 24x^2 \leq 25 - 50x$

7. $x^3 + 17x \leq 10x^2 - 28$

8. $x^3 - x^2 > 16x + 20$

9. $7x^2 + x^3 \geq 9x + 63$

10. $x^4 + 100 < 29x^2$

11. $x^4 + x^3 + 30 > x + 31x^2$

12. $x^4 + 3x^3 + 70 \leq 71x^2 + 3x$

Example 4: Determine if the inequality $x^2 + 27 < x$ has a solution.

Step 1: Write the inequality in standard form.

To write the inequality $x^2 + 27 < x$ in standard form, the terms on the left side of the inequality should be in order from the highest degree to the lowest degree and the right side of the inequality should be 0. The inequality can be converted to standard form as follows.

$$x^2 + 27 - x < x - x$$
$$x^2 - x + 27 < 0$$

Step 2: Graph the left side of the inequality.

Now, the function $f(x) = x^2 - x + 27$ should be graphed as follows.

Calculator:
On the TI-83/84 graphing calculator, this graph can be produced by performing the following steps.

a. Press $\boxed{Y=}$.

b. Enter X^2 – X + 27 after $Y_1 =$.

c. Press $\boxed{\text{WINDOW}}$.

d. Enter the following values.

Xmin = –10
Xmax = 10
Xscl = 2
Ymin = –18
Ymax = 81
Yscl = 9
Xres = 1

e. Press $\boxed{\text{GRAPH}}$.

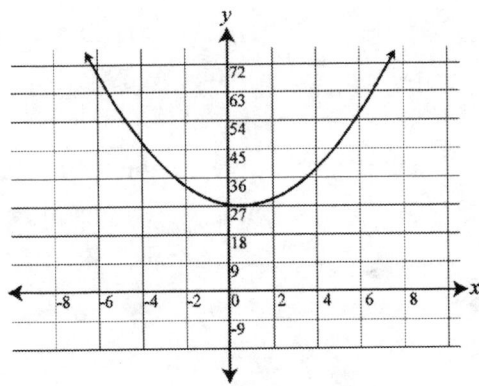

Step 3: Determine if any part of the graph satisfies the inequality.

Since the inequality in standard form $x^2 - x + 27 < 0$, the graph must be below the x-axis to satisfy the inequality. Because the graph is always above the x-axis, it never satisfies the inequality, so the inequality does not have a solution.

Determine if each of the following inequalities has a solution.

1. $x^2 + 30 < -5x$

2. $x^2 + 21 < 10x$

3. $2x - x^2 \geq 15$

4. $6x - x^2 > -40$

5. $22 + x^2 > -3x$

6. $-x^2 \leq 4x + 18$

7.3   Solving Exponential Inequalities Analytically

To solve an exponential inequality analytically, the base and exponent must first be isolated on one side of the inequality. Once the base and exponent are isolated, the logarithm of both sides of the inequality should be taken. The logarithm with base e is called the **natural logarithm**, denoted by $\ln x$. $\log_e x = \ln x$. The logarithm property $\log_e(c^d) = d\log_e c$ can be used with natural logarithms, so that it's possible to solve for the variable in the exponent. Finally, the solution to the inequality can be written in interval notation.

Example 5:   Solve the inequality $6^x < 380$ for x. Write your answer in interval notation and round to two decimal places.

Step 1:   Take the natural logarithm of both sides of the inequality.

Since the base and exponent are already isolated on one side of the inequality, the natural logarithm of both sides of the inequality should be taken as follows.

$\ln(6^x) < \ln 380$

Step 2:   Use the property $\log_e(c^d) = d\log_e c$.

$\ln(6^x) = \log_e(6^x) = x\log_e 6 = x \cdot \ln 6$

$x\ln 6 < \ln 380$

Step 3:   Solve the inequality for x.

Finally, the inequality can be solved for x as follows.

$x \cdot \ln 6 < \ln 380$

$x \cdot 1.7918 < 5.9402$

$\dfrac{x \cdot 1.7918}{1.7918} < \dfrac{5.9402}{1.7918}$

$x < \dfrac{5.9402}{1.7918}$

$x < 3.32$

Step 4:   Write the answer in interval notation.

Therefore, the solution to the inequality in interval notation is .$(-\infty, 3.32)$.

Solve each of the following inequalities for x. Write each answer in interval notation and round to two decimal places.

1. $2^x > 101$

2. $10^x < 489$

3. $\left(\frac{3}{4}\right)^x \le 68$

4. $8^x \ge 724$

5. $7^x < 97$

6. $\left(\frac{1}{5}\right)^x > 243$

Example 6: Solve the inequality $(-5)\left(7^{x-3}\right) - 45 \geq -630$ for x. Write your answer in interval notation and round to two decimal places.

Step 1: Isolate the base and exponent.

To isolate the base and exponent on one side of the inequality, first 45 must be added to both sides of the inequality and then both sides of the inequality must be divided by -5 as follows.

$$(-5)\left(7^{x-3}\right) \geq -585$$

$$\frac{(-5)\left(7^{x-3}\right)}{-5} \leq \frac{-585}{-5}$$

$$7^{x-3} \leq 117$$

Step 2: Take the natural logarithm of both sides of the inequality.

Since the base and exponent are now isolated on one side of the inequality, the natural logarithm of both sides of the inequality should be taken as follows.

$$\ln\left(7^{x-3}\right) \leq \ln 117$$

Step 3: Use the property $\log_e\left(c^d\right) = d\log_e c$.

$$\ln\left(7^{x-3}\right) = \log_e\left(7^{x-3}\right) = (x-3)\log_e 7 = (x-3) \cdot \ln 7$$

$$(x-3)\ln 7 \leq \ln 117$$

Step 4: Solve the inequality for x.

$$(x-3) \cdot 1.9459 \leq 4.7622$$

$$\frac{(x-3) \cdot 1.9459}{1.9459} \leq \frac{4.7622}{1.9459}$$

$$x - 3 \leq \frac{4.7622}{1.9459}$$

$$x - 3 \leq 2.45$$
$$x - 3 + 3 \leq 2.45 + 3$$
$$x \leq 5.45$$

Step 5: Write the answer in interval notation.

Therefore, the solution to the inequality in interval notation is $(-\infty, 5.45]$.

Solve each of the following inequalities for x. Write each answer in interval notation and round to two decimal places.

1. $(8)\left(9^{x+1}\right) + 34 \leq 202$

2. $(3)\left(11^{x+4}\right) - 81 \geq 72$

3. $\left(\frac{2}{7}\right)\left(5^{x-2}\right) - 76 < 38$

4. $(-9)\left(2^{x-5}\right) - 98 > -458$

5. $(4)\left(6^{2x+7}\right) + 183 \geq 199$

6. $\left(-\frac{3}{10}\right)\left(4^{3x-1}\right) + 221 \leq 122$

Example 7:   Solve the inequality $8e^{x+6} - 25 > 383$ for x. Write your answer in interval notation and round to two decimal places.

Step 1:   Isolate the base and exponent.

To isolate the base and exponent on one side of the inequality, first 25 must be added to both sides of the inequality and then both sides of the inequality must be divided by 8 as follows.

$$8e^{x+6} - 25 + 25 > 383 + 25$$
$$8e^{x+6} > 408$$

$$\frac{8e^{x+6}}{8} > \frac{408}{8}$$

$$e^{x+6} > 51$$

Step 2:   Take the natural logarithm of both sides of the inequality.

Since the base and exponent are now isolated on one side of the inequality, the natural logarithm of both sides of the inequality should be taken as follows.

$$\ln\left(e^{x+6}\right) > \ln(51)$$
$$x + 6 > \ln(51)$$

Step 3:   Solve the inequality for x.

Finally, the inequality can be solved for x as follows.

$$x + 6 - 6 > \ln(51) - 6$$
$$x > \ln(51) - 6$$
$$x > 3.93 - 6$$
$$x > -2.07$$

Step 4:   Write the answer in interval notation.

Therefore, the solution to the inequality in interval notation is $(-2.07, \infty)$.

Solve each of the following inequalities for x. Write each answer in interval notation and round to two decimal places.

7. $5e^{x-12} + 60 < 95$

8. $3e^{x+8} - 132 > 12$

9. $\dfrac{e^{x-3}}{2} - 309 \geq 13$

10. $-7e^{4x+3} + 54 \leq -163$

11. $17e^{\frac{x}{5}} - 13 > 89$

12. $\dfrac{e^{2x-7}}{6} + 123 < 652$

7.4   Solving Exponential Inequalities Graphically

There are two ways to solve an exponential inequality graphically.

1) Graph the left side of the inequality, graph the right side of the inequality on the same coordinate grid, and determine the regions of the graph that satisfy the inequality.

2) Move all the terms of the exponential inequality to the left side so that the right side is 0, graph the left side, and determine the regions of the graph that satisfy the inequality.

With either approach, if no region of the graph satisfies the inequality, then the inequality has no solution.

Example 8:   Solve the inequality $(7)(4^x) + 10 > 38$ graphically. Write the solution in interval notation.

Step 1:   Graph the left side of the inequality.

First, the function $f(x) = (7)(4^x) + 10$ should be graphed as follows.

> **Calculator:**
> On the TI-83/84 graphing calculator, this graph can be produced by performing the following steps.
>
> a. Press $\boxed{Y=}$.
>
> b. Enter 7*4^X + 10 after $Y_1 =$.
>
> c. Press $\boxed{\text{WINDOW}}$.
>
> d. Enter the following values.
>
> Xmin = −6
> Xmax = 4
> Xscl = 1
> Ymin = −20
> Ymax = 80
> Yscl = 10
> Xres = 1
>
> e. Press $\boxed{\text{GRAPH}}$.

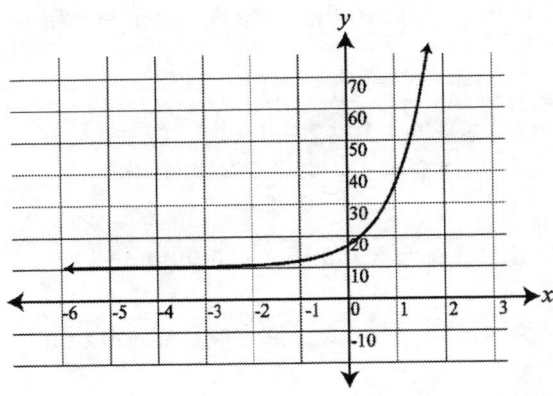

Step 2:   Graph the right side of the inequality on the same coordinate grid.

Next, the function $g(x) = 38$ should be graphed on the same coordinate grid as follows.

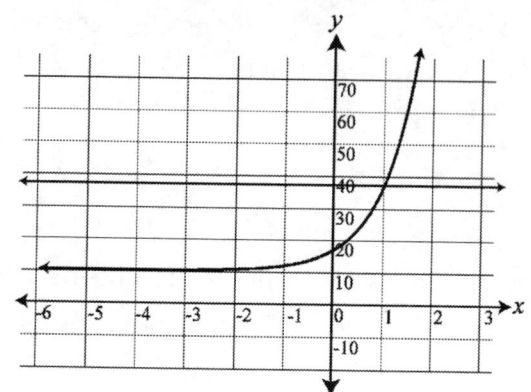

Calculator:

On the TI-83/84 graphing calculator, this graph can be produced by performing the following steps.

a. Press Y =.

b. Enter 7*4^X + 10 after Y_1 = and 38 after Y_2.

c. Press WINDOW.

d. Enter the values from Step 1.

e. Press GRAPH.

Step 3:   Determine the x-coordinate of the point where the graphs intersect.

Calculator:

On the TI-83/84 graphing calculator, the point of intersection of the graphs can be found by performing the following steps.

a. Press 2ND TRACE.

b. Press 5.

c. Press ENTER twice.

d..Move the cursor as close as possible to the point of intersection.

e. Press ENTER.

The graphs intersect at the point $(1, 38)$, so the x-coordinate is 1.

Step 4:   Determine the regions of the graph that satisfy the inequality.

Since the inequality is $(7)(4^x) + 10 > 38$, the regions of the graph that satisfy the inequality are those in which the graph of $f(x) = (7)(4^x) + 10$ is above the graph of $g(x) = 38$. This occurs when x is greater than 1.

Step 5:   Write the solution to the inequality in interval notation.

Therefore, the solution to the inequality in interval notation is $(1, \infty)$.

Example 9:   Solve the inequality $(-5)(2^x) + 52 \leq 12$ graphically. Write the solution in interval notation.

Step 1:   Move all the terms to the left side of the inequality.

When all the terms of $(-5)(2^x) + 52 \leq 12$ are moved to the left side of the inequality, the right side of the inequality becomes 0 as follows.

$$(-5)(2^x) + 52 \leq 12$$

$$(-5)(2^x) + 52 - 12 \leq 12 - 12$$

$$(-5)(2^x) + 40 \leq 0$$

Step 2:   Graph the left side of the inequality.

Now, the function $f(x) = (-5)(2^x) + 40$ should be graphed as follows.

Calculator:

On the TI-83/84 graphing calculator, this graph can be produced by performing the following steps.

a. Press $\boxed{Y=}$.

b. Enter $-5*2^\wedge X + 40$ after $Y_1 =$.

c. Press $\boxed{WINDOW}$.

d. Enter the following values.

Xmin = –5
Xmax = 5
Xscl = 1
Ymin = –40
Ymax = 60
Yscl = 10
Xres = 1

e. Press $\boxed{GRAPH}$.

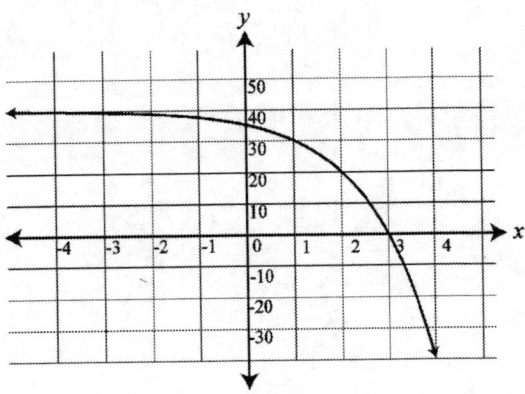

Step 3:   Determine the x-coordinate of the point where the graph crosses the x-axis.

> **Calculator:**
> On the TI-83/84 graphing calculator, the point of intersection of the graphs can be found by performing the following steps.
>
> a. Press $\boxed{\text{2ND}}$ $\boxed{\text{TRACE}}$.
>
> b. Press $\boxed{2}$.
>
> c. Move the cursor to the left of the point where the graph crosses the x-axis.
>
> d. Press $\boxed{\text{ENTER}}$.
>
> e. Move the cursor to the right of the point where the graph crosses the x-axis.
>
> f. Press $\boxed{\text{ENTER}}$.

The graph crosses the x-axis at the point $(3, 0)$, so the x-coordinate is 3.

Step 4:   Determine the regions of the graph that satisfy when all terms are moved to the left side, the regions of the graph that satisfy the inequality are those in which the graph of $f(x) = (-5)(2^x) + 40$ is at or below the x-axis. This occurs when x is greater than or equal to 3.

Step 5:   Write the solution to the inequality in interval notation.

Therefore, the solution to the inequality is interval notation is $[3, \infty)$.

Solve each of the following inequalities graphically. Write each solution in interval notation.

1. $(2)(5^x) - 29 < 21$

2. $(-8)(3^x) + 17 > -7$

3. $\left(\frac{1}{4}\right)(8^x) - 56 \geq 72$

4. $(21)(7^x) + 9 \leq 12$

5. $(3)(2^{x-4}) - 96 > 96$

6. $\left(-\frac{2}{3}\right)(6^{x+1}) + 31 < 7$

7. $(45)(9^x) + 14 \geq 19$

8. $(-32)(4^x) - 29 \leq -31$

9. $(11)(5^x) - 6 < 49$

10. $\left(\frac{1}{4}\right)(6^x) + 22 > 76$

11. $\left(\frac{4}{9}\right)(3^{x-2}) + 7 \leq 43$

12. $(7)(8^{2x+3}) - 50 \geq 6$

Example 10:   Determine if the inequality $(-4)(3^{x+5}) - 19 \geq 89$ has a solution.

Step 1:   Move all the terms to the left side of the inequality.

When all the terms of $(-4)(3^{x+5}) - 19 \geq 89$ are moved to the left side of the inequality, the right side of the inequality becomes 0 as follows.
$$(-4)(3^{x+5}) - 19 \geq 89$$
$$(-4)(3^{x+5}) - 108 \geq 0$$

Step 2:   Graph the left side of the inequality.

Now, the function $f(x) = (-4)(3^{x+5}) - 108$ should be graphed as follows.

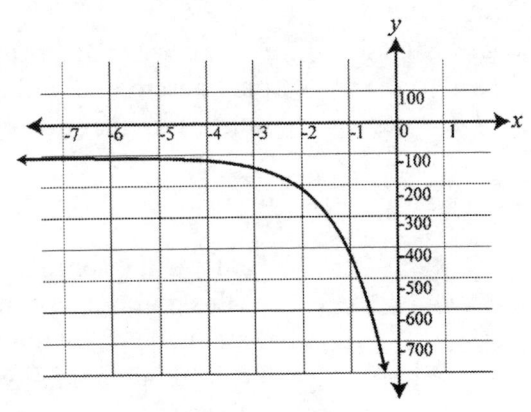

Calculator:
On the TI-83/84 graphing calculator, this graph can be produced by performing the following steps.

a. Press $\boxed{Y =}$.

b. Enter $-4*3^\wedge(X + 5) - 108$ after $Y_1 =$.

c. Press $\boxed{\text{WINDOW}}$.

d. Enter the following values.

Xmin = –8
Xmax = 2
Xscl = 1
Ymin = –800
Ymax = 200
Yscl = 100
Xres = 1

e. Press $\boxed{\text{GRAPH}}$.

Step 3:   Determine if any part of the graph satisfies the inequality.

Since the inequality is $(-4)(3^{x+5}) - 108 \geq 0$ when all the terms are moved to the left side, the graph must be at or above the x-axis to satisfy the inequality. Because the graph is always below the x-axis, it never satisfies the inequality, so the inequality does not have a solution.

Determine if each of the following inequalities has a solution.

1. $(12)(5^x) + 76 \leq 16$

2. $\left(-\frac{1}{2}\right)(4^x) + 90 \geq 82$

3. $\left(\frac{2}{3}\right)(9^x) - 60 > -6$

4. $(15)(2^{x-7}) - 13 < -73$

5. $(-18)(6^{x+3}) + 28 \geq 25$

6. $\left(\frac{3}{4}\right)(8^{3x-1}) + 55 \leq 7$

7.5   Solving Logarithmic Inequalities Analytically

Logarithmic inequalities can often be solved analytically just by using the fact that another way of writing $\log_a x > y$ is $x > a^y$, or by using the property that says if $\log_a c > \log_a d$, then $c > d$. Both of these statements hold true regardless of the inequality sign. Also, when solving a logarithmic inequality, it's necessary to take into account the fact that it's only possible to find the logarithm of a positive number. Once the solution to a logarithmic inequality is found, it can then be written in interval notation.

Example 11:   Write the solution to the inequality $\log_6 (x + 10) \leq 3$ in interval notation.

Step 1:   Use the fact that another way of writing $\log_a x \leq y$ is $x \leq a^y$.
Since another way of writing $\log_a x \leq y$ is $x \leq a^y$, the inequality $\log_6 (x + 10) \leq 3$ can also be written as $x + 10 \leq 6^3$, or $x + 10 \leq 216$.

Step 2:   Solve for x.
$x + 10 \leq 216$
$x \leq 206$

Step 3:   Also, because the original inequality was $\log_6 (x + 10) \leq 3$, and because it's only possible to find the logarithm of a positive number, the expression $x + 10$ must be greater than 0.
$x + 10 > 0$
$x > -10$

Step 4:   Combine the solutions.
At this point, it's been determined that $x \leq 206$ and $x > -10$. Since both of these inequalities must be true, they must be combined to find the solution to the original inequality $\log_6 (x + 10) \leq 3$. Only values of x that are greater than -10 and less than or equal to 206 satisfy both $x \leq 206$ and $x > -10$, so the solution to the original inequality $\log_6 (x + 10) \leq 3$ is $-10 < x \leq 206$.

Step 5:   Write the answer in interval notation.
Therefore, the solution to the inequality in interval notation is $(-10, 206]$.

Write the solution to each of the following inequalities in interval notation.

1. $\log_5 (x - 9) > 3$

2. $\log_2 (x + 20) \geq 5$

3. $\log_7 (2x - 5) \geq 2$

4. $\log (3x - 8) > 1$

5. $\log_3 (4x - 3) > 4$

6. $\log_4 \left(x + \frac{1}{2} \right) \geq -1$

7. $\log_8 (x + 5) \leq 1$

8. $\log_{12} (x - 3) < 2$

9. $\log_3 (2x - 1) < 5$

10. $\log (5x + 150) \leq 3$

11. $\log_2 \left(x - \frac{3}{8} \right) \leq -2$

12. $\log_7 \left(\frac{x}{6} + 2 \right) < 1$

Example 12:   Write the solution to the inequality $\log_5 (3x - 9) \geq \log_5 (2x - 4)$ in interval notation.

Step 1:   Use the property that says if $\log_a c \geq \log_a d$, then $c \geq d$.
If $\log_a c \geq \log_a d$, then $c \geq d$, so this means that since
$\log_5 (3x - 9) \geq \log_5 (2x - 4)$, $3x - 9 \geq 2x - 4$.

Step 2:   Solve for x.
$3x - 9 \geq 2x - 4$
$x \geq 5$

Step 3:   Take into account the fact that it's only possible to find the logarithm of a positive number.

Also, because the original inequality was $\log_5 (3x - 9) \geq \log_5 (2x - 4)$, and because it's only possible to find the logarithm of a positive number, the expressions $3x - 9$ and $2x - 4$ must be greater than 0. This means that the inequalities $3x - 9 > 0$ and $2x - 4 > 0$ should also be solved for x.

Step 4:   Solve for x.
$3x - 9 > 0$
$x > 3$

Step 5:   Solve for x.
$2x - 4 > 0$
$x > 2$

Step 6:   Combine the solutions.
At this point, it's been determined that $x \geq 5$, $x > 3$, and $x > 2$. Since all three of these inequalities must be true, they must be combined to find the solution to the original inequality $\log_5 (3x - 9) \geq \log_5 (2x - 4)$. Only values of x that are greater than or equal to 5 satisfy $x \geq 5$, $x > 3$, and $x > 2$, so the solution to the original inequality $\log_5 (3x - 9) \geq \log_5 (2x - 4)$ is $x \geq 5$.

Step 7:   Write the answer in interval notation.
Therefore, the solution to the inequality in interval notation is $[5, \infty)$.

Write the solution to each of the following inequalities in interval notation.

1. $\log_9 (4x + 8) \geq \log_9 (5x + 1)$

2. $\log_2 (12x - 6) \leq \log_2 (10x + 8)$

3. $\log (x + 20) > \log (3x + 14)$

4. $\ln (6x - 12) < \ln (7x - 21)$

5. $\log_{18} (8x - 2) \geq \log_{18} (6x - 3)$

6. $\log_7 (9x + 4) \leq \log_7 (8x + 9)$

7.6   Solving Logarithmic Inequalities Graphically

Just like solving logarithmic equations, there are two ways to solve a logarithmic inequalities graphically.

1) Graph the left side of the inequality, graph the right side of the inequality, and determine the regions of the graph that satisfy the inequality.

2) Move all the terms of the logarithmic inequality to the left side, graph the left side, and determine the regions of the graph that satisfy the inequality.

With either approach, if no region of the graph satisfies the inequality, then the inequality has no solution.

Remember to use the logarithmic property $\log_a c = \dfrac{\log c}{\log a}$ when graphing.

Example 13:   Solve the inequality $\log (x + 6) > 1$ graphically. Write the solution in interval notation.

Step 1:   Graph the left side of the inequality.

First, the function $f(x) = \log(x + 6)$ should be graphed as follows.

Calculator:
a. Press $\boxed{Y =}$.

b. Enter $\log(X + 6)$ after $Y_1 =$.

c. Press $\boxed{\text{WINDOW}}$.

d. Enter the following values.
Xmin = −10
Xmax = 10
Xscl = 2
Ymin = −2.5
Ymax = 2.5
Yscl = 0.5
Xres = 1

e. Press $\boxed{\text{GRAPH}}$.

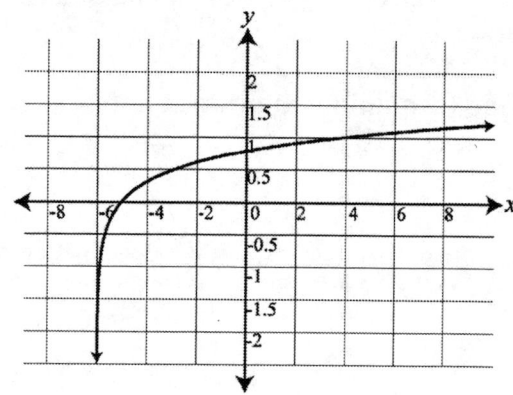

Step 2:   Graph the right side of the inequality on the same coordinate grid.

Next, the function $g(x) = 1$ should be graphed on the same coordinate grid as follows.

Calculator:
a. Press $\boxed{Y=}$.

b. Enter 1 after $Y_2 =$.

c. Press $\boxed{\text{GRAPH}}$.

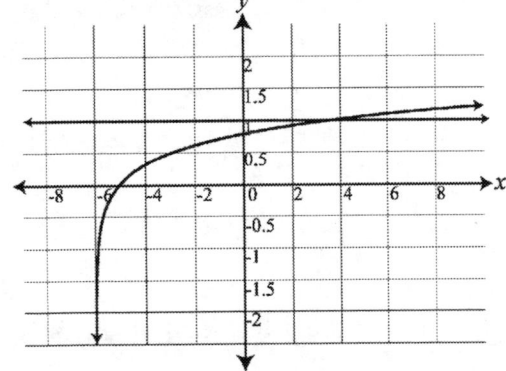

Step 3:   Determine the x-coordinate of the point where the graphs intersect.

Calculator:
a. Press $\boxed{\text{2ND}}$ $\boxed{\text{TRACE}}$.

b. Press $\boxed{5}$.

c. Press $\boxed{\text{ENTER}}$ twice.

d. Move the cursor as close as possible to the point of intersection.

e. Press $\boxed{\text{ENTER}}$.

The graphs intersect at the point $(4, 1)$, so the x-coordinate is 4.

Step 4:   Determine the regions of the graph that satisfy the inequality.

Since the inequality is $\log(x + 6) > 1$, the regions of the graph that satisfy the inequality are those in which the graph of $f(x) = \log(x + 6)$ is above the graph the $g(x) = 1$. This occurs when x is greater than 4.

Step 5:   Write the solution to the inequality in interval notation.

Therefore, the solution to the inequality in interval notation is $(4, \infty)$.

Solve each of the following inequalities graphically. Write each solution in interval notation.

1. $\log(x - 2) \geq 1$

2. $\log_3(x + 5) > 2$

3. $\log_2(x - 8) > 4$

4. $\log_4(x + 50) \geq 3$

5. $\log_5(x + 19) \geq 2$

6. $\log_7(x - 10) > 0$

Example 14:   Solve the inequality $\log_2 (x - 4) \leq 2$ graphically. Write the solution in interval notation.

Step 1:   Since $\log_a c = \dfrac{\log c}{\log a}$, the inequality $\log_2 (x - 4) \leq 2$ can be rewritten as

$$\frac{\log (x - 4)}{\log (2)} \leq 2.$$

Step 2:   Move all the terms to the left side of the inequality.

$$\frac{\log (x - 4)}{\log (2)} - 2 \leq 0$$

Step 3:   Graph the left side of the inequality.

Now, the function $f(x) = \dfrac{\log (x - 4)}{\log (2)} - 2$ should be graphed as follows.

Calculator:

a. Press $\boxed{Y =}$.

b. Enter $\log(X - 4)/\log(2) - 2$ after Y_1.

c. Press $\boxed{\text{WINDOW}}$.

d. Enter the following values.
Xmin = –4
Xmax = 16
Xscl = 2
Ymin = –6
Ymax = 4
Yscl = 1
Xres = 1

e. Press $\boxed{\text{GRAPH}}$.

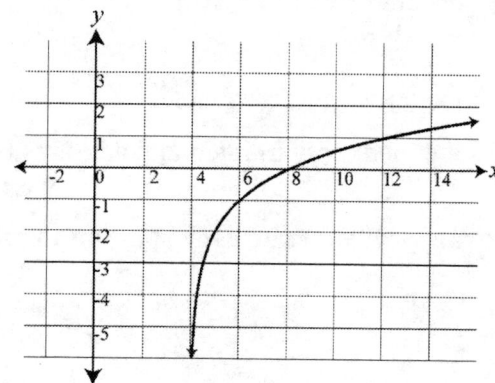

Step 4:   Determine the x-coordinate of the point where the graph crosses the x-axis.

> **Calculator:**
> a. Press $\boxed{\text{2ND}}$ $\boxed{\text{TRACE}}$.
>
> b. Press $\boxed{2}$.
>
> c. Move the cursor to the left of the point where the graph crosses the x-axis.
>
> d. Press $\boxed{\text{ENTER}}$.
>
> e. Move the cursor to the left of the point where the graph crosses the x-axis.
>
> f. Press $\boxed{\text{ENTER}}$ twice.

The graph crosses the x-axis at the point $(8, 0)$, so the x-coordinate is 8.

Step 5:   Determine the regions of the graph that satisfy the inequality.

Since the inequality is $\dfrac{\log(x - 4)}{\log(2)} - 2 \leq 0$ when all the terms are moved to the left side, the regions of the graph that satisfy the inequality are those in which the graph of $f(x) = \dfrac{\log(x - 4)}{\log(2)} - 2$ is at or below the x-axis.

This occurs when x is greater than 4 and less than or equal to 8.

Step 6:   Write the solution to the inequality in interval notation.

Therefore, the solution to the inequality in interval notation is $(4, 8]$.

Solve each of the following inequalities graphically. Write each solution in interval notation.

1. $\log_{15}(x + 6) \leq 1$ 3. $\log_3(x + 12) < 3$ 5. $\log_9(x - 7) < 1$

2. $\log_4(x - 3) < 2$ 4. $\log_2(x - 1) \leq 5$ 6. $\log_5(x + 4) \leq 2$

Example 15:   Determine if the inequality $\log_3 (x^2 + 16) < 1$ has a solution.

Step 1:   Since $\log_a c = \dfrac{\log c}{\log a}$, the inequality $\log_3 (x^2 + 16) < 1$ can be rewritten as

$\dfrac{\log (x^2 + 16)}{\log (3)} < 1.$

Step 2:   Move all the terms to the left side of the inequality.

$\dfrac{\log (x^2 + 16)}{\log (3)} - 1 < 0$

Step 3:   Graph the left side of the inequality.

Now, the function $f(x) = \dfrac{\log (x^2 + 16)}{\log (3)} - 1$ should be graphed as follows.

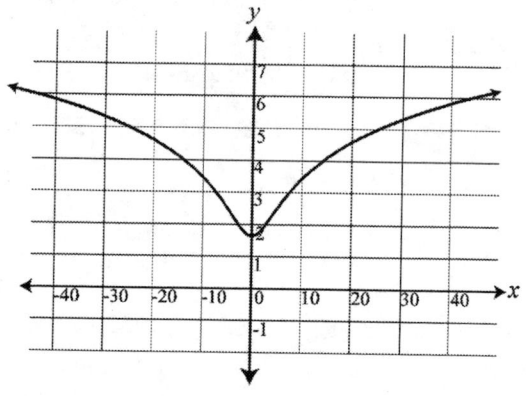

Calculator:

a. Press $\boxed{Y=}$.

b. Enter $\log(X^{\wedge}2 + 16)/\log(3) - 1$ after Y_1 .

c. Press $\boxed{\text{WINDOW}}$.

d. Enter the following values.

Xmin = −50
Xmax = 50
Xscl = 10
Ymin = −2
Ymax = 8
Yscl = 1
Xres = 1

e. Press $\boxed{\text{GRAPH}}$.

Step 4:   Determine if any part of the graph satisfies the inequality.

Since the inequality is $\dfrac{\log (x^2 + 16)}{\log (3)} - 1 < 0$ when all the terms are moved to the left side, the graph must be below the x-axis, it never satisfies the inequality, so the inequality does not have a solution.

Determine if each of the following inequalities has a solution.

1. $\log_9 (x^2 + 100) > 2$

2. $\log_2 (x^2 + 78) < 6$

3. $-\log (x^2 + 32) \geq 1$

4. $-\log_4 (x^2 + 86) \leq 3$

5. $\log_7 (x^2 + 25) < 2$

6. $\log (x^2 + 244) > 3$

Chapter 7 Review

Write the solution to each of the following inequalities in interval notation.

1. $x^2 - 3x \leq 54$

2. $x^2 > 7 - 6x$

3. $x^2 \geq 7x + 60$

4. $x^3 + 5x^2 < 34x + 80$

5. $x^3 + 8x^2 \leq 72 + 9x$

6. $x^3 + 12x^2 > 36 - 23x$

To solve the polynomial inequality $x^3 - 2x^2 \geq 8x$, both sides of the inequality were graphed as shown below.

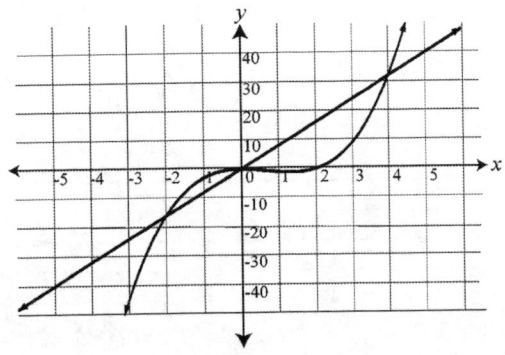

7. What two functions were graphed?

8. What is the solution to the polynomial inequality $x^3 - 2x^2 \geq 8x$ in interval notation?

To solve the polynomial inequality $x^4 + 32 < x^3 + 18x^2 - 16x$, the inequality was written in standard form and then the left side of the inequality was graphed as shown below.

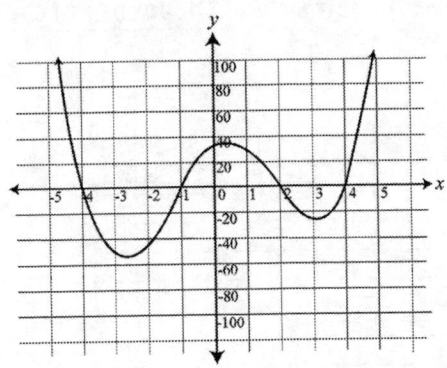

9. What function was graphed?

10. What is the solution to the polynomial inequality $x^4 + 32 < x^3 + 18x^2 - 16x$ in interval notation?

Determine if each of the following inequalities has a solution.

11. $x^2 + 7x < -30$

12. $x^2 + 22 \geq 5x$

Solve each of the following inequalities for x. Write each answer in interval notation and round to two decimal places.

13. $9^x \leq 204$

14. $\left(\frac{1}{7}\right)^x > 73$

15. $\left(\frac{1}{9}\right)\left(3^{x+6}\right) + 89 < 192$

16. $(8)\left(2^{x-10}\right) - 225 \geq 39$

17. $11e^{7x-1} + 52 > 151$

18. $\dfrac{e^{2x+5}}{8} - 27 \leq 96$

To solve the exponential inequality $(5)\left(8^{x+4}\right) - 111 > 209$**, both sides of the inequality were graphed as shown below.**

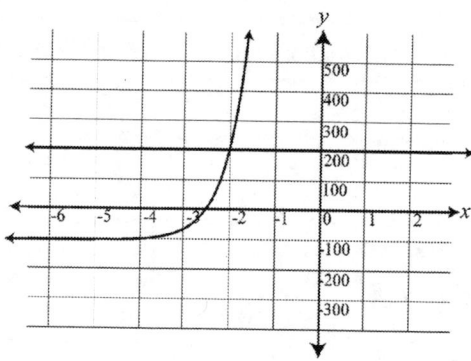

19. What two functions were graphed?

20. What is the solution to the exponential inequality $(5)\left(8^{x+4}\right) - 111 > 209$ in interval notation?

To solve the exponential inequality $(70)\left(10^{x+8}\right) + 16 \leq 23$**, all the terms were moved to the left side of the inequality and then the left side of the inequality was graphed as shown below.**

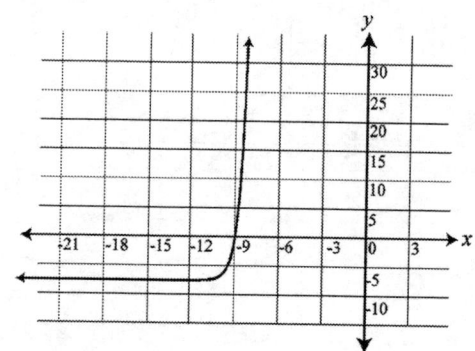

21. What function was graphed?

22. What is the solution to the exponential inequality $(70)\left(10^{x+8}\right) + 16 \leq 23$ in interval notation?

Determine if each of the following inequalities has a solution.

23. $\left(-\frac{1}{9}\right)\left(6^{5x+2}\right) + 100 \geq 96$

24. $(25)\left(2^{x-6}\right) + 290 < 90$

Write the solution to each of the following inequalities in interval notation.

25. $\log_6 (x - 11) < 2$

28. $\log \left(x + \frac{1}{5} \right) \geq -1$

26. $\log_5 (x + 3) \leq 3$

29. $\log_4 (5x + 25) \geq \log_4 (6x + 18)$

27. $\log_{14} (x - 2) > 1$

30. $\log_7 (12x - 1) < \log_7 (8x + 11)$

To solve the logarithmic inequality $\log_2 (x + 5) < 3$, both sides of the inequality were graphed as shown below.

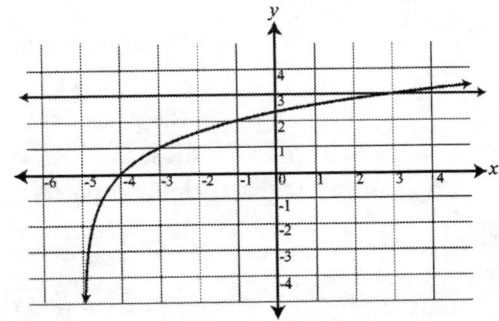

31. What two functions were graphed?

32. What is the solution to the logarithmic inequality $\log_2 (x + 5) < 3$ in interval notation?

To solve the logarithmic inequality $\log_{18} (x + 12) \geq 1$, all the terms were moved to the left side of the inequality and then the left side of the inequality was graphed as shown below.

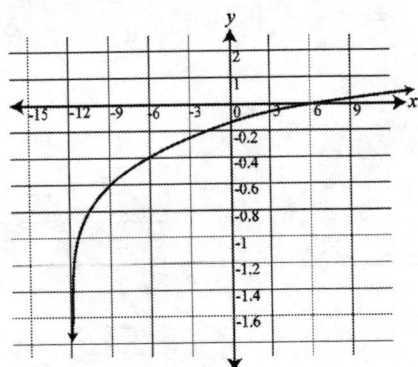

33. What function was graphed?

34. What is the solution to the logarithmic inequality $\log_{18} (x + 12) \geq 1$ in interval notation?

Determine if each of the following inequalities has a solution.

35. $\log_4 (x^2 + 288) < 4$

36. $-\log_5 (x^2 + 79) \leq 2$

Chapter 7 Test

1. How many regions should a number line be broken into test the inequality $x^2 > 6x + 40$?

   A. 2
   B. 3
   C. 4
   D. 5

2. The solution to an inequality is $-5 \leq x \leq 4$. What is the solution written in interval notation?

   A. $(-5, 4)$
   B. $[-5, 4]$
   C. $(4, -5)$
   D. $[4, -5]$

3. What is the solution to the inequality $x^2 + 36 > 15x$ written in interval notation?

   A. $(3, 12)$
   B. $[3, 12]$
   C. $(-\infty, 3) \cup (12, \infty)$
   D. $(-\infty, 3] \cup [12, \infty)$

4. The graph of the function $f(x) = x^3 - x^2 - 5x - 3$ is shown below.

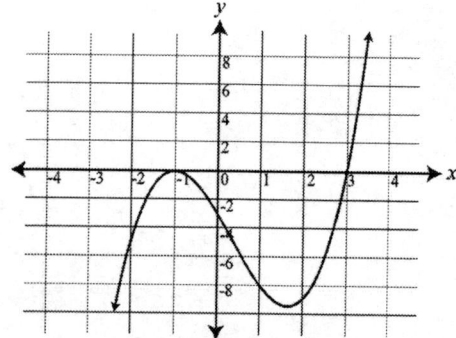

What is the solution to the inequality $x^3 - x^2 - 5x - 3 > 0$ in interval notation?

   A. $(3, \infty)$
   B. $[3, \infty)$
   C. $(-\infty, 3)$
   D. $(-\infty, 3]$

5. Which of these inequalities does not have a solution?

   A. $x^2 - 100 \leq 0$
   B. $x^2 - 100 \geq 0$
   C. $x^2 + 100 \leq 0$
   D. $x^2 + 100 \geq 0$

6. To solve the inequality $x^3 - x^2 - 81x + 81 \geq 0$, the function $f(x) = x^3 - x^2 - 81x + 81$ was graphed on a coordinate grid. Which of these regions of the coordinate grid satisfy the inequality?

   A. Those where $f(x)$ is above the x-axis.
   B. Those where $f(x)$ is at or above the x-axis.
   C. Those where $f(x)$ is below the x-axis.
   D. Those where $f(x)$ is at or below the x-axis.

7. The graphs of the functions $f(x) = x^3 + 2x^2$ and $g(x) = 36x + 72$ are shown below.

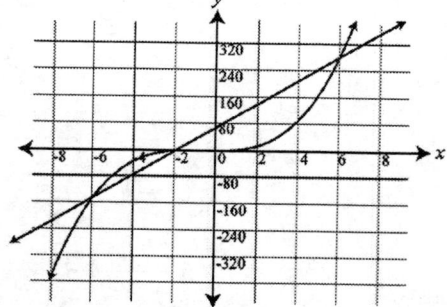

What is the solution to the inequality $x^3 + 2x^2 \leq 36x + 72$ in interval notation?

   A. $(-6, -2) \cup (6, \infty)$
   B. $[-6, -2] \cup [6, \infty]$
   C. $(-\infty, -6) \cup (-2, 6)$
   D. $(-\infty, -6] \cup [-2, 6]$

8. What is the solution to the inequality $12^x \geq 998$ rounded to two decimal places and written in interval notation?

   A. $(0.36, \infty)$

   B. $[0.36, \infty)$

   C. $(2.78, \infty)$

   D. $[2.78, \infty)$

9. When solving an exponential inequality analytically, which of these steps should be performed first?

   A. Take the logarithm of both sides of the inequality.

   B. Isolate the base and exponent.

   C. Write the answer in interval notation.

   D. Use the property $\log_a (c^d) = d \log_a c$.

10. What is the solution to the inequality $7e^{5x+2} - 45 > 11$ rounded to two decimal places and written in interval notation?

    A. $(0.02, \infty)$

    B. $[0.02, \infty)$

    C. $(0.82, \infty)$

    D. $[0.82, \infty)$

11. In the course of solving the inequality $\left(\frac{1}{2}\right)(4^x) + 17 \leq 61$, which of these inequalities will be arrived at?

    A. $4^x \leq 22$

    B. $4^x \leq 39$

    C. $4^x \leq 88$

    D. $4^x \leq 156$

12. The graph of the function $f(x) = (5)(9^{x-1}) - 15$ is shown below.

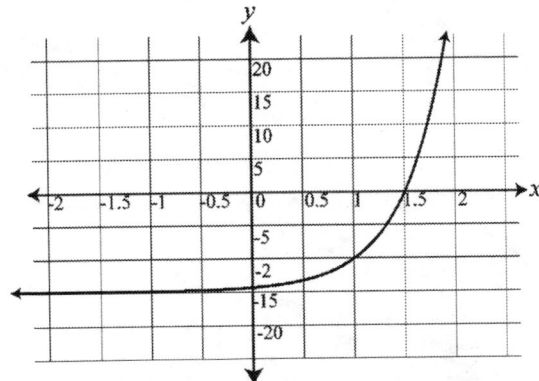

What is the solution to the inequality $(5)(9^{x-1}) - 15 \geq 0$ in interval notation?

    A. $(-15, \infty)$

    B. $[-15, \infty)$

    C. $(1.52, \infty)$

    D. $[1.5, \infty)$

13. To solve the inequality $\left(\frac{5}{8}\right)(4^x) + 18 \leq 28$, all the terms were moved to the left side of the inequality and the left side was graphed. Which if these functions was graphed?

    A. $f(x) = \left(\frac{5}{8}\right)(4^x) - 46$

    B. $f(x) = \left(\frac{5}{8}\right)(4^x) - 10$

    C. $f(x) = \left(\frac{5}{8}\right)(4^x) + 10$

    D. $f(x) = \left(\frac{5}{8}\right)(4^x) + 46$

14. To solve the inequality $(8)(6^{2x+3}) - 19 < 29$, the functions $f(x) = (8)(6^{2x+3}) - 19$ and $g(x) = 29$ were graphed. When is the graph of $g(x)$ below the graph of $f(x)$?

    A. When x is less than -1

    B. When x is less than or equal to -1

    C. When x is greater than -1

    D. When x is greater than or equal to -1

15. The graphs of the functions $f(x) = (20)(3^{x+3}) - 17$ and $g(x) = 43$ are shown below.

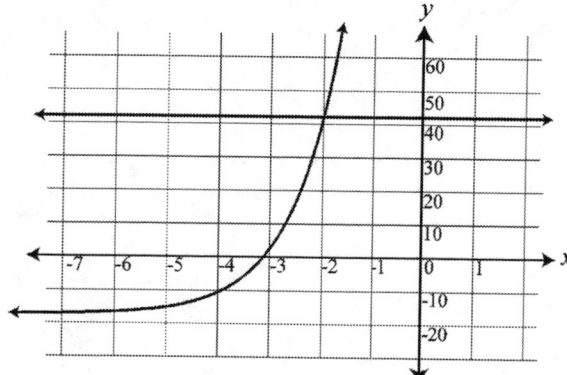

What is the solution to the inequality $(20)(3^{x+3}) - 17 > 43$ in interval notation?

A. $(-\infty, -2)$
B. $[-\infty, -2)$
C. $(-2, \infty)$
D. $[-2, \infty)$

16. What two inequalities must be combined to get the solution to the inequality $\log_9(x - 14) \leq 2$?

A. $x - 14 \leq 81$ and $x - 14 < 0$
B. $x - 14 \leq 81$ and $x - 14 > 0$
C. $x - 14 \geq 81$ and $x - 14 < 0$
D. $x - 14 \geq 81$ and $x - 14 > 0$

17. What is the solution to the inequality $\log_6(x + 106) \leq 3$ in interval notation?

A. $(-110, 106)$
B. $(-110, 106]$
C. $(-106, 110)$
D. $(-106, 110]$

18. If $\log_a x > y$, then which of these expressions must be less than x?

A. a^x
B. a^y
C. x^a
D. y^a

19. What is the solution to the inequality $\log_7(2x + 4) > \log_7(x - 3)$ in interval notation?

A. $(-7, \infty)$
B. $(-2, \infty)$
C. $(0, \infty)$
D. $(3, \infty)$

20. To solve the inequality $\log_4(3x - 1) \geq \log_4(x + 9)$, the functions $f(x) = \log_4(3x - 1)$ and $g(x) = \log_4(x + 9)$ were graphed on the same coordinate grid. Which of these regions of the coordinate grid satisfy the inequality?

A. Those where $f(x)$ is above $g(x)$.
B. Those where $f(x)$ is at or above $g(x)$.
C. Those where $f(x)$ is below $g(x)$.
D. Those where $f(x)$ is at or below $g(x)$.

21. How can the inequality $\log_{11}(x + 3) - 2 > 0$ be rewritten so that the left side can be graphed on a graphing calculator?

A. $\dfrac{\log(x + 3)}{\log(2)} - 11 > 0$

B. $\dfrac{\log(2)}{\log(x + 3)} - 11 > 0$

C. $\dfrac{\log(x + 3)}{\log(11)} - 2 > 0$

D. $\dfrac{\log(11)}{\log(x + 3)} - 2 > 0$

22. Does the inequality $-\log(x^2 + 5) - 1 \leq 0$ have a solution?

A. No, because the graph of the left side is always below the x-axis.
B. No, because the graph of the left side is always above the x-axis.
C. Yes, because the graph of the left side is always below the x-axis.
D. Yes, because the graph of the left side is always above the x-axis.

Chapter 8
Graphing Polynomial Functions

This chapter covers the following Georgia Performance Standards:

| MM3A | Algebra | MM3A1a, MM3A1b, MM3A1c, MM3A1d |

8.1 Characteristics of Quadratic Functions

A function in the form $f(x) = a_n x^n + a_{n-1}x^{n-1} + ... + a_2 x^2 + a_1 x + a_0$, where the coefficients a_n, a_{n-1}, ... a_2, a_1, and a_0 are constants and the exponents are whole numbers, is called a polynomial function. Polynomial functions and their graphs have certain characteristics that include x-intercepts, y-intercepts, and zeros, relative and absolute maxima and minima, domain and range, intervals of increase and decrease, and end behavior.

Example 1: Determine the x-intercepts, y-intercept, and zeros, relative and absolute maxima and minima, domain and range, intervals of increase and decrease, and end behavior of the polynomial function $f(x) = 5x^2 - 31x + 6$. Round to the nearest tenth, when necessary.

Step 1: Graph the function.
First, the function $f(x) = 5x^2 - 31x + 6$ can be graphed as follows.

Calculator:

a. Press $\boxed{Y =}$.

b. Enter $5X\hat{\ }2 - 31X + 6$ after $Y_1 =$.

c. Press $\boxed{WINDOW}$.

d. Enter the following values.

Xmin = –3
Xmax = 9
Xscl = 1
Ymin = –50
Ymax = 50
Yscl = 10
Xres = 1

e. Press $\boxed{GRAPH}$.

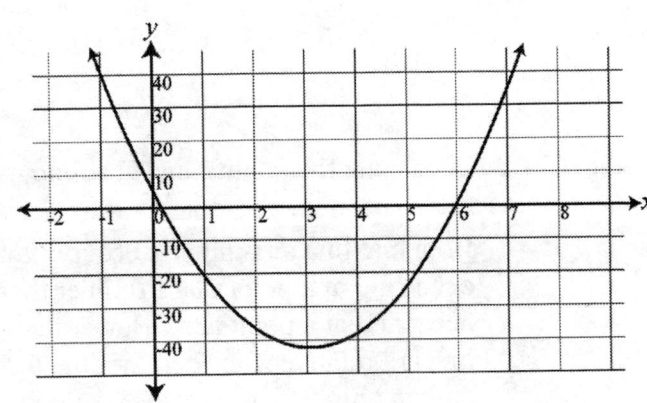

Step 2:   Determine the x-intercepts and zeros.

Both the x-intercepts and zeros can be found by determining where the graph of the function crosses the x-axis. Since the graph crosses the x-axis at the points $(0.2, 0)$ and $(6, 0)$, the x-intercepts are $(0.2, 0)$ and $(6, 0)$, and the zeros are $x = 0.2$ and 6. To find the zeros with a calculator, use the following:

From the window of the graph of the equation, continue with these steps:

a. Press ⬚2ND⬚ ⬚TRACE⬚.

b. Press ⬚2⬚.

c. Move the cursor to the left of a point where the graph crosses the x-axis.

d. Press ⬚ENTER⬚.

e. Move the cursor to the right of a point where the graph crosses the x-axis.

f. Press ⬚ENTER⬚ twice.

g. Repeat the steps above for the remaining points where the graph crosses the x-axis.

Step 3:   Determine the y-intercept.

The y-intercept can be found by determining where the graph of the function crosses the y-axis. Since the graph crosses the y-axis at the point $(0, 6)$, this is the y-intercept. Often, the easiest way to determine the y-intercept is by plugging 0 into the function for x. This is done with the function $f(x) = 5x^2 - 31x + 6$ as follows.

$$f(x) = 5x^2 - 31x + 6$$

$$f(0) = 5(0)^2 - 31(0) + 6 = 0 - 0 + 6 = 6$$

Step 4:   Determine the relative and absolute maxima and minima.

Both maxima and minima occur at turning points on the graph, with a maxima occurring at a point that is higher than all the points around it and a minima occurring at a point that is lower than all the points around it. Maxima and minima are relative if they are not the highest or lowest points on the graph, but only the highest or lowest of the points "relative' to them. Maxima and minima are absolute if they are the highest or lowest points on the graph. If a graph has only one turning point, it has no relative maxima or minima. This is the case with the graph of the function $f(x) = 5x^2 - 31x + 6$.

Also, since both the left-hand side and the right-hand side of the graph of the function $f(x) = 5x^2 - 31x + 6$ rise toward positive infinity, the graph has no absolute maximum, since there is no single point that is higher than all the others. To find the minimum on the graph of the equation, use the following steps.

From the window of the graph of the equation, continue with these steps:

a. Press ⟨2ND⟩ ⟨TRACE⟩.

b. Press ⟨3⟩.

c. Move the cursor to the left of the minimum point.

d. Press ⟨ENTER⟩.

e. Move the cursor to the right of the minimum point.

f. Press ⟨ENTER⟩ twice.

Therefore, the graph has only an absolute minimum, which occurs at the point $(3.0\overline{9}, -42.05)$, since this point is lower than all the others on the graph. This means that the absolute minimum (only the y-value) is -42.05.

Step 5:   Determine the domain and range.

The domain of the function $f(x) = 5x^2 - 31x + 6$ is all possible values of y. Since x can be any number, the domain is all real numbers, or $(-\infty, \infty)$, and since y can be any number greater than or equal to the absolute minimum, the range is all real numbers greater than or equal to -42.05, or $[-42.05, \infty)$.

Step 6:   Determine the intervals of increase and decrease.

The intervals of increase are those where the graph is going up, and the intervals of decrease are those where the graph is going down. Since the graph of the function $f(x) = 5x^2 - 31x + 6$ is decreasing (y-values are decreasing) from $x = -\infty$ to $x = 3.0\overline{9}$, its interval of decrease is $(-\infty, 3.0\overline{9}]$, and since it's increasing from $x = 3.0\overline{9}$ to $x = \infty$, its interval of increase is $[3.0\overline{9}, \infty)$.

Step 7:   Determine the end behavior.

Finally, since y approaches infinity as x approaches both negative infinity and positive infinity, the end behavior is $f(x) \to \infty$ as $x \to \pm\infty$.

Determine each of the following for the polynomial function $f(x) = 3x^2 + x - 4$. Round to the nearest tenth, when necessary.

1. x-intercepts, y-intercepts, and zeros

2. relative and absolute maxima and minima

3. domain and range

4. intervals of increase and decrease

5. end behavior

8.2   Characteristics of Cubic Functions

Example 2:   Determine the x-intercepts, y-intercept, and zeros, relative and absolute maxima and minima, domain and range, intervals of increase and decrease, and end behavior of the polynomial function $f(x) = x^3 - 4x^2 - 4x + 16$. Round to the nearest tenth, when necessary.

Step 1:   The function $f(x) = x^3 - 4x^2 - 4x + 16$ can graphed as follows.

Calculator:

a. Press $\boxed{Y=}$.

b. Enter X^3 − 4X^2 − 4X + 16 after $Y_1 =$.

c. Press $\boxed{\text{WINDOW}}$.

d. Enter the following values.

Xmin = −4
Xmax = 6
Xscl = 1
Ymin = −25
Ymax = 25
Yscl = 5
Xres = 1

e. Press $\boxed{\text{GRAPH}}$.

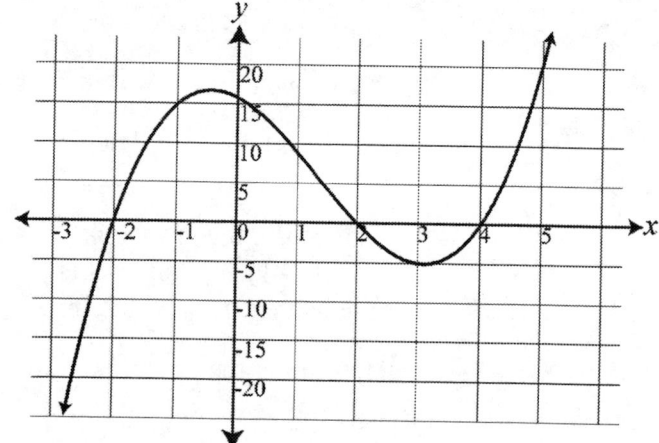

Step 2:   Determine the x-intercepts and zeros.

Both the x-intercepts and zeros can be found by determining where the graph of the function crosses the x-axis. Since the graph crosses the x-axis at the points $(-2, 0)$, $(2, 0)$, and $(4, 0)$, the x-intercepts are $(-2, 0)$, $(2, 0)$, and $(4, 0)$, and the zeros are $x = -2, 2,$ and 4.

From the window of the graph of the equation, continue with these steps:

a. Press $\boxed{\text{2ND}}$ $\boxed{\text{TRACE}}$.

b. Press $\boxed{2}$.

c. Move the cursor to the left of a point where the graph crosses the x-axis.

d. Press $\boxed{\text{ENTER}}$.

e. Move the cursor to the right of a point where the graph crosses the x-axis.

f. Press $\boxed{\text{ENTER}}$ twice.

g. Repeat the steps above for the remaining points where the graph crosses the x-axis.

Step 3:   Determine the y-intercept.

The y-intercept can be found by determining where the graph of the function crosses the y-axis. Since the graph crosses the y-axis at the point $(0, 16)$, this is the y-intercept. Often, the easiest way to determine the y-intercept is by plugging 0 into the function for x. This is done with the function

$f(x) = x^3 - 4x^2 - 4x + 16$ as follows.

$f(x) = x^3 - 4x^2 - 4x + 16$

$f(0) = (0)^3 - 4(0)^2 - 4(0) + 16 = 0 - 0 - 0 + 16 = 16$

Step 4:   Determine the relative and absolute maxima and minima.

Since the right-hand side of the graph of the function $f(x) = x^3 - 4x^2 - 4x + 16$ rises toward infinity, the graph has no absolute maximum, since there is no single point that is higher than all the others. However, it does have a relative maximum, which occurs at the point $(-0.4, 16.9)$, since this point is higher than all the points "relative" to it. This means that the relative maximum is 16.9.

From the window of the graph of the equation, continue with these steps:

a. Press [2ND] [TRACE].

b. Press [4].

c. Move the cursor to the left of the maximum point.

d. Press [ENTER].

e. Move the cursor to the right of the maximum point.

f. Press [ENTER] twice.

Since the left-hand side of the graph of the function $f(x) = x^3 - 4x^2 - 4x + 16$ falls toward negative infinity, the graph has no absolute minimum, since there is no single point that is lower than all the others. However, it does have a relative minimum, which occurs at the point $(3.1, -5.0)$, since this point is lower than all the points "relative" to it. This means that the relative minimum is -5.0.

From the window of the graph of the equation, continue with these steps:

a. Press [2ND] [TRACE].

b. Press [3].

c. Move the cursor to the left of the minimum point.

d. Press [ENTER].

e. Move the cursor to the right of the minimum point.

f. Press [ENTER] twice.

Step 5:   Determine the domain and range.

The domain of the function $f(x) = x^3 - 4x^2 - 4x + 16$ is all possible values of x, and the range is all possible values of y. Since x can be any number, the domain is all real numbers, or $(-\infty, \infty)$, and since y can be any number, the range is all real numbers, or $(-\infty, \infty)$.

Step 6:   Determine the intervals of increase and decrease.

The intervals of increase are those where the graph is going up, and the intervals of decrease are those where the graph is going down. Since the graph of the function $f(x) = x^3 - 4x^2 - 4x + 16$ is increasing from $x = -\infty$ to $x = -0.4$ and from $x = 3.1$ to $x = \infty$, its intervals of increase are $(-\infty, -0.4]$ and $[3.1, \infty]$, and since it's decreasing from $x = -0.4$ to $x = 3.1$, its interval of decrease is $[0.4, 3.1]$.

Step 7:   Determine the end behavior.

Finally, since y approaches negative infinity as x approaches negative infinity, and since y approaches infinity as x approaches infinity, the end behavior is $f(x) \to -\infty$ as $x \to -\infty$ and $f(x) \to \infty$ as $x \to \infty$.

Determine each of the following for the polynomial function $f(x) = -x^3 + 9x^2 - 23x + 15$. Round to the nearest tenth, when necessary.

1. x-intercepts, y-intercept, and zeros

2. relative and absolute maxima and minima

3. domain and range

4. intervals of increase and decrease

5. end behavior

8.3   Characteristics of Other Polynomial Functions

Example 3:   Determine the x-intercepts, y-intercept, and zeros, relative and absolute maxima and minima, domain and range, intervals of increase and decrease, and end behavior of the polynomial function $f(x) = -4x^4 - 4x^3 + 9x^2 + x - 2$. Round to the nearest tenth, when necessary.

Step 1:   Graph the function.

First, the function $f(x) = -4x^4 - 4x^3 + 9x^2 + x - 2$ can be graphed as follows.

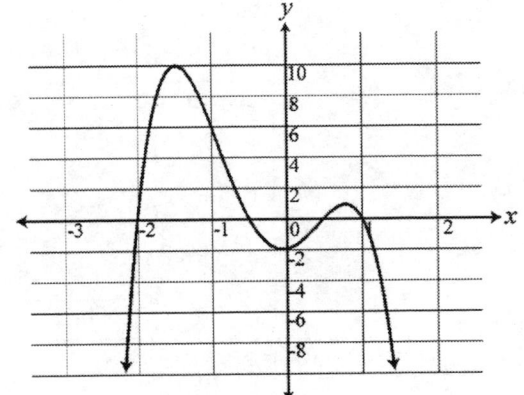

Calculator:
a. Press $\boxed{Y=}$.

b. Enter $-4X^4 - 4X^3 + 9X^2 + X - 2$ after $Y_1 =$.

c. Press $\boxed{\text{WINDOW}}$.

d. Enter the following values.

Xmin = –4
Xmax = 3
Xscl = 1
Ymin = –10
Ymax = 12
Yscl = 2
Xres = 1

e. Press $\boxed{\text{GRAPH}}$.

Step 2:   Determine the x-intercepts and zeros.
Both the x-intercepts and zeros can be found by determining where the graph of the function crosses the x-axis. Since the graph crosses the x-axis at the points $(-2, 0)$, $(-0.5, 0)$, $(0.5, 0)$, and $(1, 0)$, the x-intercepts are $(-2, 0)$, $(-0.5, 0)$, $(0.5, 0)$, and $(1, 0)$, and the zeros are $x = -2, -0.5, 0.5$, and 1.

From the window of the graph of the equation, continue with these steps:

a. Press $\boxed{\text{2ND}}$ $\boxed{\text{TRACE}}$.

b. Press $\boxed{2}$.

c. Move the cursor to the left of a point where the graph crosses the x-axis.

d. Press $\boxed{\text{ENTER}}$.

e. Move the cursor to the right of a point where the graph crosses the x-axis.

f. Press $\boxed{\text{ENTER}}$ twice.

g. Repeat the steps above for the remaining points where the graph crosses the x-axis.

Step 3:   Determine the y-intercept.

The y-intercept can be found by determining where the graph of the function crosses the y-axis. Since the graph crosses the y-axis at the point $(0, -2)$, this is the y-intercept. Often, the easiest way to determine the y-intercept is by plugging 0 into the function for x. This is done with the function

$$f(x) = -4x^4 - 4x^3 + 9x^2 + x - 2$$
$$f(0) = -4(0)^4 - 4(0)^3 + 9(0)^2 + 0 - 2 = 0 - 0 + 0 + 0 - 2 = -2$$

Step 4:   Determine the relative and absolute maxima and minima.

The graph has two maxima, which occur at the points $(-1.5, 10.0)$ and $(0.8, 0.9)$. The absolute maximum is 10.0, since the point $(-1.5, 10.0)$ is higher than all the other points on the graph, and 0.9 is a relative maximum, since the point $(0.8, 0.9)$ is higher than all the points "relative" to it.

From the window of the graph of the equation, continue with these steps:

a. Press $\boxed{\text{2ND}}$ $\boxed{\text{TRACE}}$.

b. Press $\boxed{4}$.

c. Move the cursor to the left of the maximum point.

d. Press $\boxed{\text{ENTER}}$.

e. Move the cursor to the right of the maximum point.

f. Press $\boxed{\text{ENTER}}$ twice.

g. Repeat the steps above for the remaining maximum points.

Since both the left-hand side and the right-hand side of the graph of the function $f(x) = -4x^4 - 4x^3 + 9x^2 + x - 2$ fall toward negative infinity, the graph has no absolute minimum, since there is no single point that is lower than all the others. However, it does have a relative minimum, which occurs at the point $(0, -2)$, since this point is lower than all the points "relative" to it. This means that the relative minimum is -2.

From the window of the graph of the equation, continue with these steps:

a. Press $\boxed{\text{2ND}}$ $\boxed{\text{TRACE}}$.

b. Press $\boxed{3}$.

c. Move the cursor to the left of the minimum point.

d. Press $\boxed{\text{ENTER}}$.

e. Move the cursor to the right of the minimum point.

f. Press $\boxed{\text{ENTER}}$ twice.

g. Repeat the steps above for the remaining minimum points.

Step 5:   Determine the domain and range.

The domain of the function $f(x) = -4x^4 - 4x^3 + 9x^2 + x - 2$ is all possible values of x, and the range is all possible values of y. Since x can be any number, the domain is all real numbers, or $(-\infty, \infty)$, and since y can be any number less than or equal to the absolute maximum, the range is all real numbers less than or equal to 10.0, or $(-\infty, 10.0]$.

Step 6:   Determine the intervals of increase and decrease.

The intervals of increase are those where the graph is going up, and the intervals of decrease are those where the graph is going down. Since the graph of the function $f(x) = -4x^4 - 4x^3 + 9x^2 + x - 2$ is increasing from $x = -\infty$ to $x = 1.5$ and from $x = 0$ to $x = 0.8$, its intervals of increase are $(-\infty, 1.5]$ and $[0, 0.8]$, and since it's decreasing from $x = -1.5$ to $x = 0$ and from $x = 0.8$ to $x = \infty$, its intervals of decrease are $[-1.5, 0]$ and $[0.8, \infty)$.

Step 7:   Determine the end behavior.

Finally, since y approaches negative infinity as x approaches both negative infinity and positive infinity, the end behavior is $f(x) \to -\infty$ as $x \to \pm\infty$.

Determine each of the following for the polynomial function $f(x) = -x^4 - x^3 + 7x^2 + x - 6$. Round to the nearest tenth, when necessary.

1. x-intercepts, y-intercept, and zeros

2. relative and absolute maxima and minima

3. domain and range

4. intervals of increase and decrease

5. end behavior

8.4   Graph Transformations of $f(x) = ax^n$

The graph of the function $f(x) = ax^n$ can be transformed by vertically stretching or compressing it, reflecting it across the x-axis, horizontally stretching or compressing it, reflecting it across the y-axis, translating it right or left, and translating it up or down. Each of these transformations is described in the table below.

Transformation	Transformed Function	Conditions
Vertical Stretch or Compression	$f(x) = c \cdot ax^n$	c is a constant that has an absolute value greater than 1 for a vertical stretch and an absolute value less than 1 for a vertical compression
Reflection Across the x-axis	$f(x) = c \cdot ax^n$	c is a constant that is negative
Horizontal Stretch or Compression	$f(x) = a(d \cdot x)^n$	d is a constant that has an absolute value less than 1 for a horizontal stretch and an absolute value greater than 1 for a horizontal compression
Reflection Across the y-axis	$f(x) = a(d \cdot x)^n$	d is the constant that is negative
Translation Right or Left	$f(x) = a(x - h)^n$	h is a constant that is positive for a translation right and negative for a translation left
Translation Up or Down	$f(x) = ax^n + k$	k is a constant that is positive for a translation up and negative for a translation down

Example 4:   The graph of the function $f(x) = 2x^2$ was transformed to produce the graph of the function $f(x) = -6x^2$ as shown below. Determine the transformations that were applied.

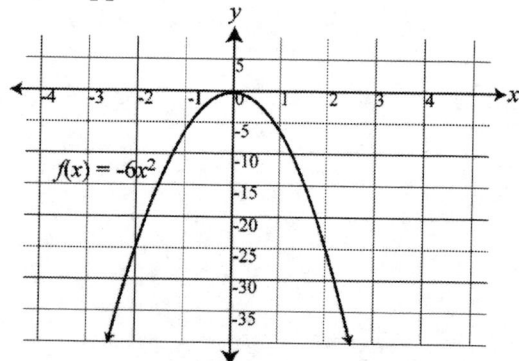

Step 1:   Check for a vertical stretch or compression.
The function $f(x) = -6x^2$ is in the form $f(x) = c \cdot 2x^2$, with the value of c being -3. Since the absolute value of -3 is greater than 1, the graph of the function $f(x) = 2x^2$ has undergone a vertical stretch.

Step 2:   Check for a reflection across the x-axis.
Also, since c is negative, the graph of the function $f(x) = 2x^2$ has undergone a reflection across the x-axis.

Example 5:   The graph of the function $f(x) = 32x^3$ was transformed by a horizontal stretch, a horizontal compression, a reflection across the y-axis, or a combination of these transformations to produce the graph of the function $f(x) = 4x^3$ as shown below. Determine the transformations that were applied.

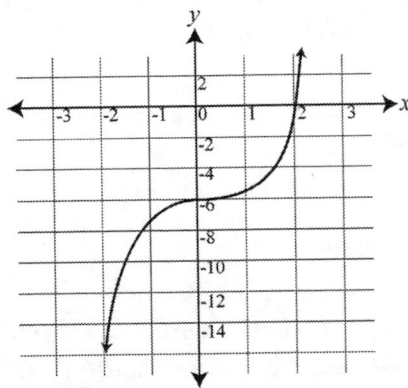

Step 1:   Check for horizontal stretch or compression.

The function $f(x) = 4x^3$ is in the form $f(x) = 32(d \cdot x)^3$, with the value of d being $\frac{1}{2}$. Since the absolute value of $\frac{1}{2}$ is less than 1, the graph of the function $f(x) = 32x^3$ has undergone a horizontal stretch.

Step 2:   Check for a reflection across the y-axis.

Also, since d is positive, the graph of the function $f(x) = 32x^3$ has not undergone a reflection across the y-axis.

Example 6:   The graph of the function $f(x) = \dfrac{x^4}{12}$ was transformed by a translation right or left, a translation up or down, or a combination of these transformations to produce the graph of the function $f(x) = \dfrac{(x-2)^4}{12} + 5$ as shown below. Determine the transformations that were applied.

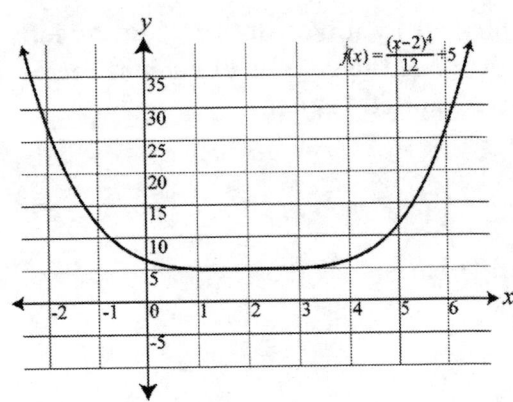

Step 1:   Check for a translation right or left.

The function $f(x) = \dfrac{(x-2)^4}{12} + 5$ is in the form $f(x) = \dfrac{(x-h)^4}{12} + k$, with the value of h being 2. Since h is positive, the graph of the function $f(x) = \dfrac{x^4}{12}$ has undergone a translation right.

Step 2:   Check for a translation up or down.

Also, the value of k is 5. Since k is positive, the graph of the function $f(x) = \dfrac{x^4}{12}$ has undergone a translation up.

The function $f(x) = 10x^3$ was transformed by a vertical stretch, a vertical compression, a reflection across the x-axis, or a combination of these to produce the graphs of each of the following functions. Determine the transformations that were applied to each.

1. $f(x) = -5x^3$

2. $f(x) = 40x^3$

3. $f(x) = -100x^3$

4. $f(x) = \dfrac{x^3}{10}$

5. $f(x) = \dfrac{25x^3}{2}$

6. $f(x) = -x^3$

The function $f(x) = x^5$ was transformed by a horizontal stretch, a horizontal compression, a reflection across the y-axis, or a combination of these to produce the graphs of each of the following functions. Determine the transformations that were applied to each.

7. $f(x) = -1024x^5$

8. $f(x) = \dfrac{7x^5}{3}$

9. $f(x) = 0.6x^5$

10. $f(x) = -1.9x^5$

11. $f(x) = -\dfrac{x^5}{243}$

12. $f(x) = 11x^5$

The function $f(x) = 3.5x^2$ was transformed by a translation right or left, a translation up or down, or a combination of these to produce the graphs of each of the following functions. Determine the transformations that were applied to each.

13. $f(x) = 3.5(x+7)^2 - 13$

14. $f(x) = 3.5(x-2.2)^2 + 8.9$

15. $f(x) = 3.5(x+10)^2$

16. $f(x) = 3.5x^2 + 475$

17. $f(x) = 3.5(x-46)^2 - 6$

18. $f(x) = 3.5(x+3)^2 + 20$

8.5   Multiplicity of Graphs of Polynomial Functions

The graph of a polynomial function is affected by the multiplicity of its real zeros, its degree, and its lead coefficient. If the multiplicity of a real zero of a polynomial function is even – that is, if the real zero occurs an even number of times – the graph of the polynomial function touches the x-axis at this value of x, and if the multiplicity is odd, the graph passes through the x-axis at this value of x. The degree, or value of the highest exponent, and the lead coefficient of the function can affect the function's graph in different ways, depending on whether the degree is even or odd and whether the lead coefficient is positive or negative. The affects of the degree and lead coefficient are summarized in the table below.

	Even Degree	Odd Degree
Positive Lead Coefficient	• The graph has an absolute minimum. • The graph has no absolute maximum. • The domain of the function is $(-\infty, \infty)$, and the range is [absolute minimum, ∞]. • The end behavior of the graph is $f(x) \to \infty$ as $x \to \pm\infty$.	• The graph has no absolute maximum or minimum. • The domain of the function is $(-\infty, \infty)$, and the range is $(-\infty, \infty)$. • The end behavior of the graph is $f(x) \to -\infty$ as $x \to -\infty$ and $f(x) \to \infty$ as $x \to \infty$.
Negative Lead Coefficient	• The graph has an absolute maximum. • The graph has no absolute minimum. • The domain of the function is $(-\infty, \infty)$, and the range is $(-\infty,$ absolute maximum$]$. • The end behavior of the graph is $f(x) \to -\infty$ as $x \to \pm\infty$.	• The graph has no absolute maximum or minimum. • The domain of the function is $(-\infty, \infty)$, and the range is $(-\infty, \infty)$. • The end behavior of the graph is $f(x) \to \infty$ as $x \to -\infty$ and $f(x) \to -\infty$ as $x \to \infty$.

Example 7:   For the polynomial function graphed below, determine the function's real zeros and whether the multiplicity of each zero is even or odd. Also determine whether the graph has an absolute maximum, whether the graph has an absolute minimum, whether the domain and range of the function are both $(-\infty, \infty)$, and the end behavior of the graph. Finally, determine whether the function's lead coefficient is positive or negative, and whether the function's degree is even or odd.

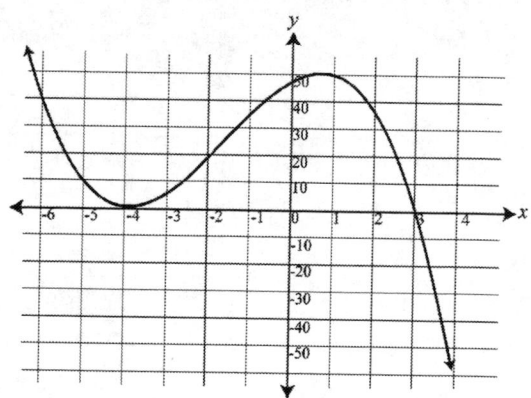

Step 1:   Determine the function's real zeros.

Since the graph of the function touches the x-axis at $x = -4$ and passes through the x-axis at $x = 3$, the function's real zeros are $x = -4$ and $x = 3$.

Step 2:   Determine whether the multiplicity of each zero is even or odd.

If the graph touches the x-axis, the multiplicity of the zero is even, and if it passes through the x-axis, the multiplicity of the zero is odd. This means that the multiplicity of the zero $x = -4$ is even and the multiplicity of the zero $x = 3$ is odd.

Step 3:   Determine whether the graph has an absolute maximum.

Since the graph extends infinitely upward and to the left, the graph does not have an absolute maximum.

Step 4:   Determine whether the graph has an absolute minimum.

Also, since the graph extends infinitely downward and to the right, the graph does not have an absolute minimum.

Step 5:   Determine whether the domain and range of the function are both $(-\infty, \infty)$.

This means that the domain of the function is $(-\infty, \infty)$, and the range is $(-\infty, \infty)$.

Step 6:   Determine the end behavior of the graph.

It also means that the end behavior of the graph is $f(x) \to \infty$ as $x \to -\infty$ and $f(x) \to -\infty$ as $x \to \infty$.

Step 7:   Determine whether the function's lead coefficient is positive or negative.

Therefore, the function's lead coefficient is negative.

Step 8:   Determine whether the function's degree is even or odd.

Also, the function's degree is odd.

For the polynomial function graphed below, determine each of the following.

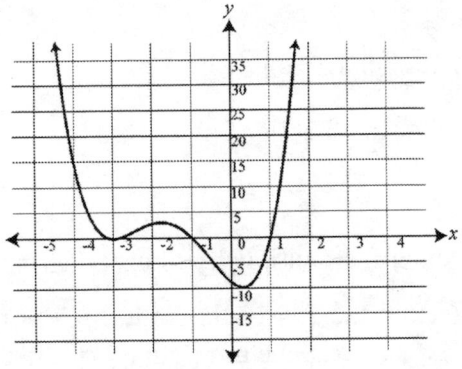

1. What are the function's real zeros?

2. Is the multiplicity of each zero even or odd?

3. Does the graph have an absolute maximum?

4. Does the graph have an absolute minimum?

5. Are the domain and range of the function both $(-\infty, \infty)$?

6. What is the end behavior?

7. Is the function's lead coefficient positive or negative?

8. Is the function's degree even or odd?

Example 8:   For the polynomial function graphed below, determine the function's real zeros and whether the multiplicity of each zero is even or odd. Also determine whether the graph has an absolute maximum, whether the graph has an absolute minimum, whether the domain and range of the function are both $(-\infty, \infty)$, and the end behavior of the graph. Finally, determine whether the function's lead coefficient is positive or negative, and whether the function's degree is even or odd.

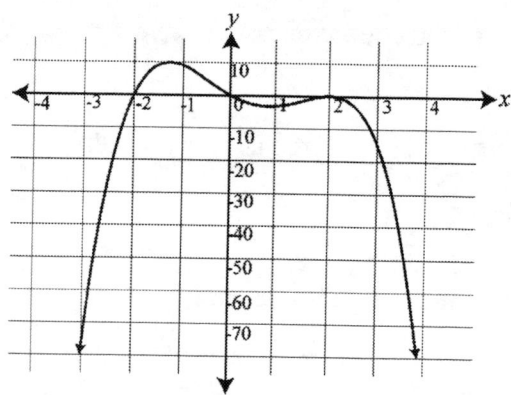

Step 1:   Determine the function's real zeros.

Since the graph of the function passes through the x-axis at $x = -2$, $x = 0$, and $x = 2$.

Step 2:   Determine whether the multiplicity of each zero is even or odd.

If the graph touches the x-axis, the multiplicity of the zero is even, and if it passes through the x-axis, the multiplicity of the zero is odd. This means that the multiplicity of the zero $x = 2$ is even and the multiplicities of the zeros $x = -2$ and $x = 0$ are odd.

Step 3:   Determine whether the graph has an absolute maximum.

Since the graph does not extend infinitely upward, the graph has an absolute maximum.

Step 4:   Determine whether the graph has an absolute minimum.

Also, since the graph extends infinitely downward and to the left and infinitely downward and to the right, the graph does not have an absolute minimum.

Step 5:   Determine whether the domain and range of the function are both $(-\infty, \infty)$.

This means that the domain of the function is $(-\infty, \infty)$, but the range is not $(-\infty, \infty)$.

Step 6:   Determine the end behavior of the graph.

It also means that the end behavior of the graph is $f(x) \to -\infty$ as $x \to \pm\infty$.

Step 7:   Determine whether the function's lead coefficient is positive or negative.

Therefore, the function's lead coefficient is negative.

Step 8:   Determine whether the function's degree is even or odd.

Also, the function's degree is even.

For the polynomial function graphed below, determine each of the following.

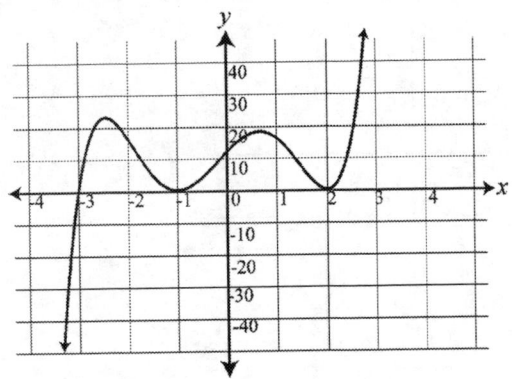

1. What are the function's real zeros?

2. Is the multiplicity of each zero even or odd?

3. Does the graph have an absolute maximum?

4. Does the graph have an absolute minimum?

5. Are the domain and range of the function both $(-\infty, \infty)$?

6. What is the end behavior?

7. Is the function's lead coefficient positive or negative?

8. Is the function's degree even or odd?

8.6   Symmetry

In Mathematics, functions can be defined as symmetrical with respect to the y-axis or the origin. To test equations for symmetry, it is helpful to remember the following:

$f(-x) = f(x)$ means the function is symmetrical with respect to the y-axis.
Being symmetrical with the y-axis means the function is **even**.

$f(-x) = -f(x)$ means the function is symmetrical with respect to the origin.
Being symmetrical with the origin means the function is **odd**.

$f(-x) \neq -f(x)$ or $f(x)$ then the function is not symmetrical.

Example 9:   Determine whether the polynomial function $f(x) = -2x^2 + 5$ graphed below is even and whether it is symmetric with respect to the y-axis.

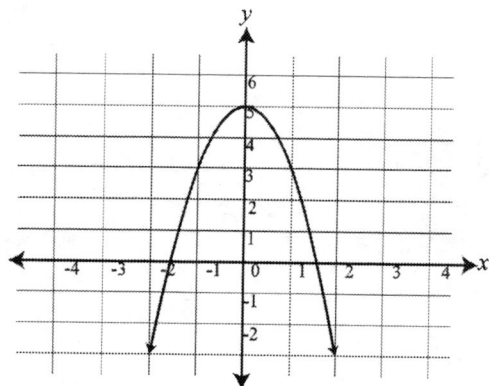

Step 1:   Determine if the function is even.

An even function is one for which $f(-x) = f(x)$. In other words, with an even function, when $-x$ is plugged in for x, the result is the original function. In this case, the function $f(x) = -2x^2 + 5$ can be checked as follows.

$f(x) = -2x^2 + 5$

$f(-x) = -2(-x)^2 + 5$

$f(-x) = -2x^2 + 5$

Since $f(x)$ and $f(-x)$ are both equal to $-2x^2 + 5$, the function is even.

Step 2:   Determine whether the function is symmetric with respect to the y-axis.

Since an even function is symmetric with respect to the y-axis, this must be the case with the function $f(x) = -2x^2 + 5$. It's also apparent from the graph that the function is mirrored about the y-axis.

Determine whether each of the following polynomial functions is even and whether each is symmetric with respect to the y-axis.

1. $f(x) = 6x^2 + 1$

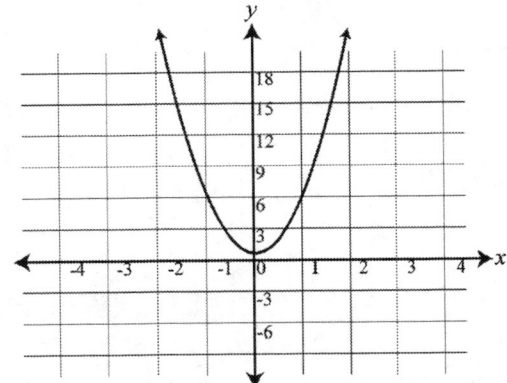

4. $f(x) = 3x^4 - 2$

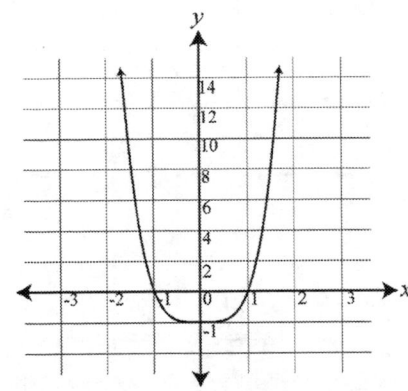

2. $f(x) = x^2 - 2x + 1$

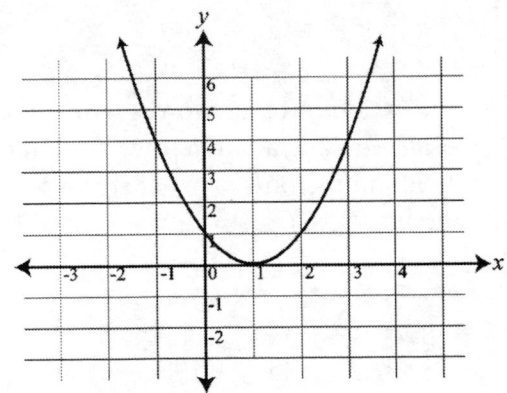

5. $f(x) = x^5 - 8$

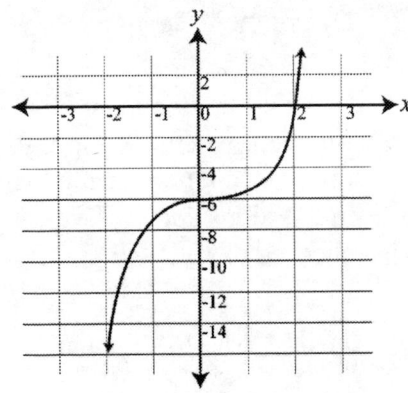

3. $f(x) = 10x^3$

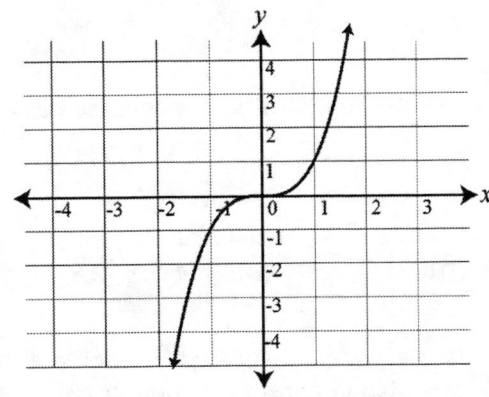

6. $f(x) = -x^6 - 12$

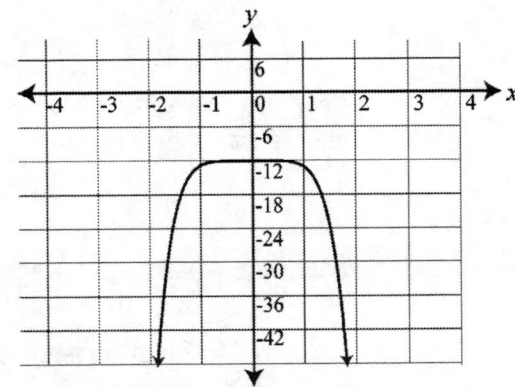

Example 10:   Determine whether the polynomial function $f(x) = 4x^3 - 9x$ graphed below is odd and whether it is symmetric with respect to the origin.

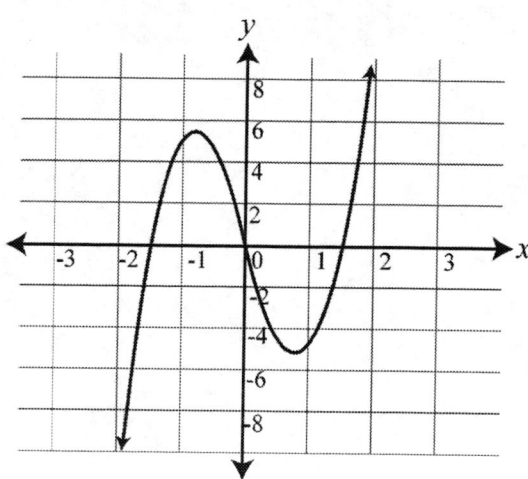

Step 1:   Determine if the function is odd.

An odd function is one for which $f(-x) = -f(x)$. In other words, with an odd function, when $-x$ is plugged in for x, the result is the additive inverse of the original function, or the original function with all the plus signs changed to minus signs and vice-versa. In this case, the function $f(x) = 4x^3 - 9x$ can be checked as follows.

$$f(x) = 4x^3 - 9x$$

$$f(-x) = 4(-x)^3 - 9(-x)$$

$$f(-x) = -4x^3 + 9x$$

Since the polynomials $4x^3 - 9x$ and $-4x^3 + 9x$ are additive inverses of each other, the function is odd.

Step 2:   Determine whether the function is symmetric with respect to the origin.

Since an odd function is symmetric with respect to the origin, this must be the case with the function $f(x) = 4x^3 - 9x$. It's also apparent from the graph that the function is mirrored about the origin.

Copyright © American Book Company

Determine whether each of the following polynomial functions is odd and whether each is symmetric with respect to the origin.

1. $f(x) = \dfrac{x^3}{2} + x$

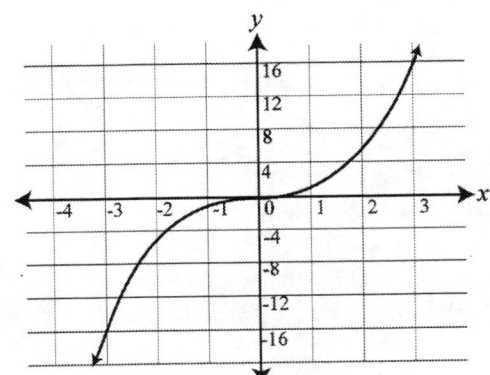

4. $f(x) = -8x^5 + 15x$

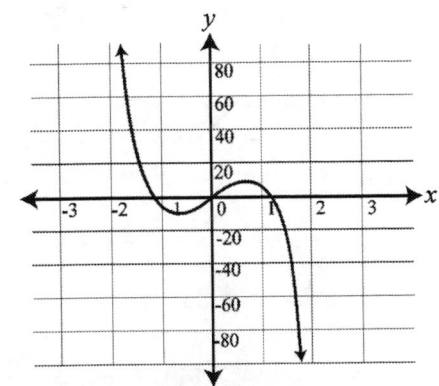

2. $f(x) = 7x^3 - 25$

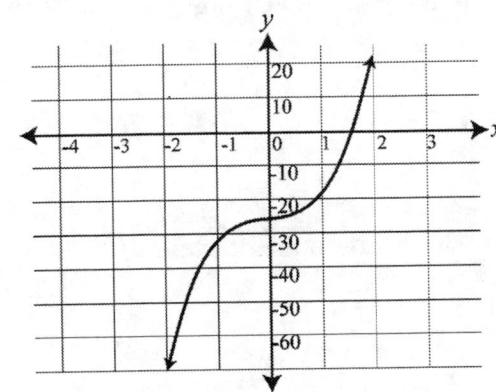

5. $f(x) = 6x^7 - 49x$

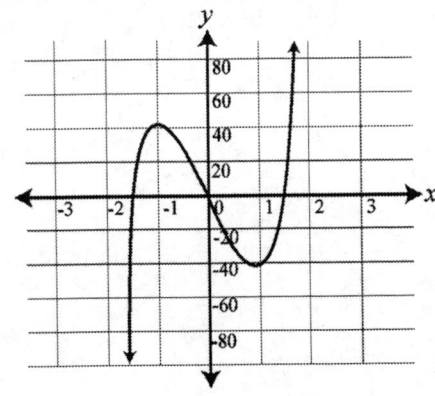

3. $f(x) = 5x^2 - 3x$

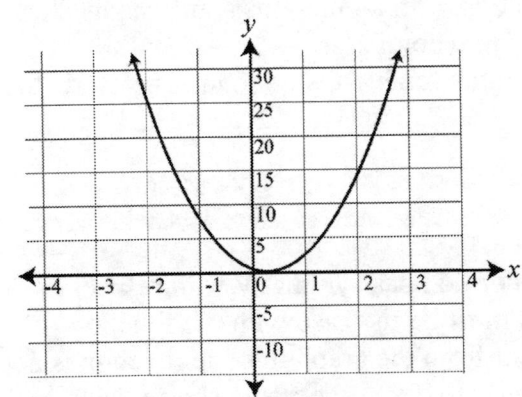

6. $f(x) = -9x^8 - 2$

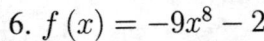
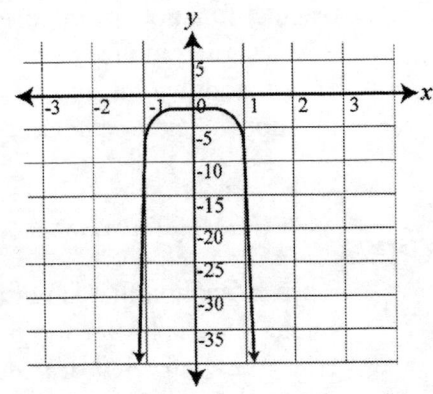

Example 11:   Determine whether the polynomial function $f(x) = x^2 + 6x + 9$ graphed below is even, odd, or neither and what its symmetries are.

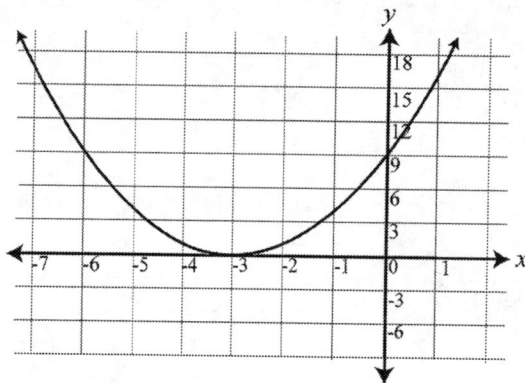

Step 1:   Determine if the function is even.

An even function is one for which $f(-x) = f(x)$. In other words, with an even function, when $-x$ is plugged in for x, the result is the original function. In this case, the function $f(x) = x^2 + 6x + 9$ can be checked as follows.

$$f(x) = x^2 + 6x + 9$$

$$f(-x) = (-x)^2 + 6(-x) + 9$$

$$f(-x) = x^2 - 6x + 9$$

Since $f(-x)$ is not equal to $x^2 + 6x + 9$, the function is not even.

Step 2:   Determine if the function is odd.

An odd function is one for which $f(-x) = -f(x)$. In other words, with an odd function, when $-x$ is plugged in for x, the result is the additive inverse of the original function, or the original function with all the plus signs changed to minus signs and vice-versa. Since the polynomials $x^2 + 6x + 9$ and $x^2 - 6x + 9$ are not additive inverses of each other, the function is not odd. Therefore, the function is neither even nor odd.

Step 3:   Determine the function's symmetries.

Since a function that is neither even nor odd has symmetry with respect to neither the y-axis nor the origin, this must be the case with the function $f(x) = x^2 + 6x + 9$. It's also apparent from the graph that the function is not mirrored about the y-axis or the origin. However, the graph shows that the function is mirrored about the line $x = -3$, so it's symmetric with respect to this line.

Determine whether each of the following polynomial functions is even, odd, or neither and what the symmetries are for each.

1. $f(x) = x^2 - 4x + 4$

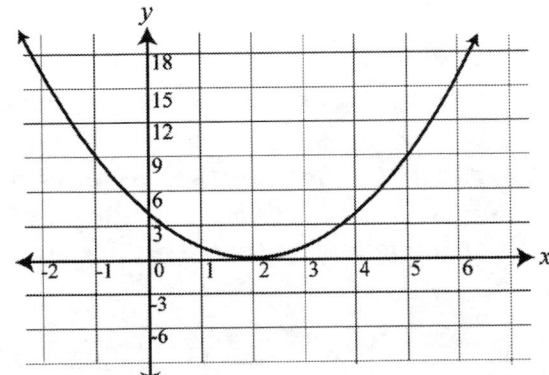

4. $f(x) = -x^3 - 3$

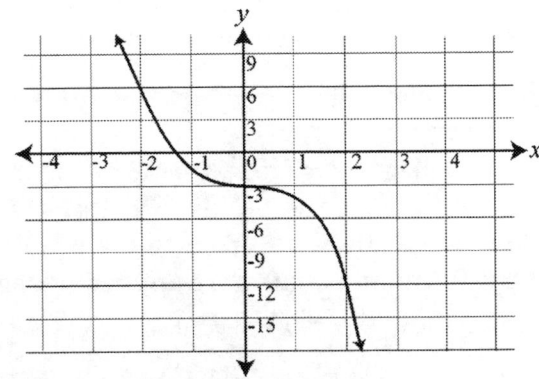

2. $f(x) = -9x^3 - 7x$

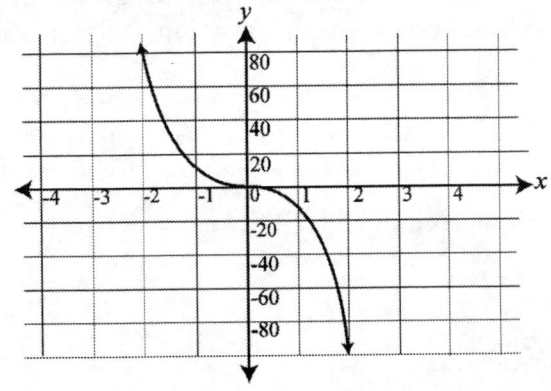

5. $f(x) = -x^2 - 8x - 16$

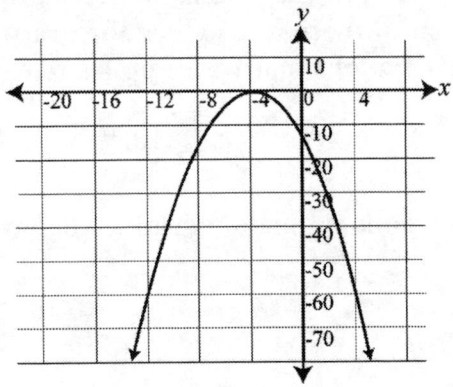

3. $f(x) = 16x^2 + 5$

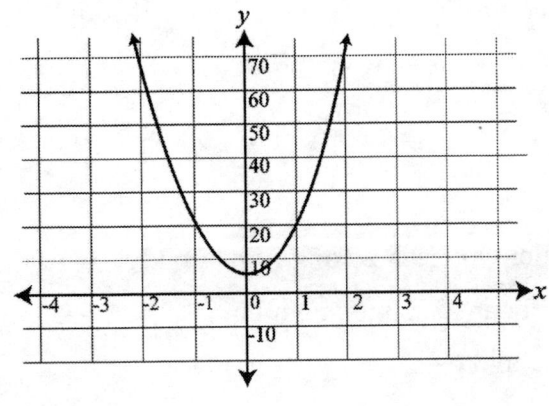

6. $f(x) = 3x^3 + 4$

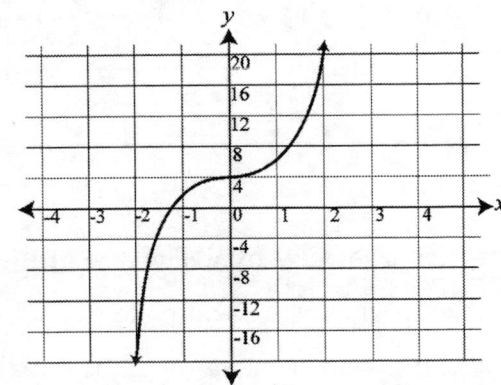

Chapter 8 Review

Determine each of the following for the polynomial function $f(x) = x^3 + 5x^2 - 8x - 12$. Round to the nearest tenth, when necessary.

1. x-intercepts, y-intercept, and zeros

2. Relative and absolute maxima and minima

3. Domain and range

4. Intervals of increase and decrease

5. End behavior

The function $f(x) = 16x^2$ was transformed by a vertical stretch, a vertical compression, a reflection across the x-axis, a translation right or left, a translation up or down, or a combination of these to produce the graphs of each of the following functions. Determine the transformations that were applied to each.

6. $f(x) = -8x^2$

7. $f(x) = 16(x-5)^2$

8. $f(x) = 48x^2 + 3$

The function $f(x) = 24x^3$ was transformed by a horizontal stretch, a horizontal compression, a reflection across the y-axis, a translation right or left, a translation up or down, or a combination of these to produce the graphs of each of the following functions. Determine the transformations that were applied to each.

9. $f(x) = 3x^3$

10. $f(x) = 192x^3 - 7$

11. $f(x) = 24(x+2)^3$

For the polynomial function graphed below, determine each of the following.

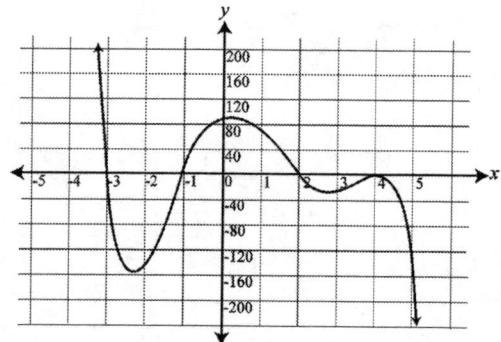

12. The function's real zeros and whether the multiplicity of each zero is even or odd

13. Whether the graph has an absolute maximum and/or an absolute minimum

14. Whether the domain and range of the function are both $(-\infty, \infty)$

15. The end behavior of the graph

16. Whether the function's lead coefficient is positive or negative

17. Whether the function's degree is even or odd

Determine whether each of the following polynomial functions is even, odd, or neither and what the symmetries are for each.

18. $f(x) = 11x^3 + 2x$

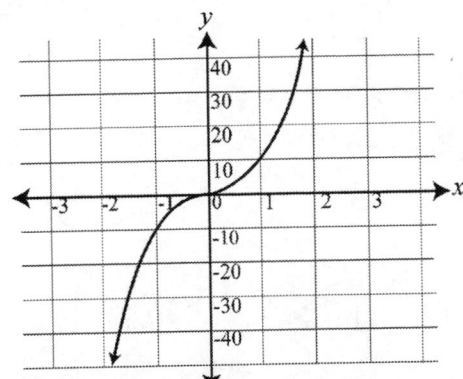

21. $f(x) = 4x^4 - 15$

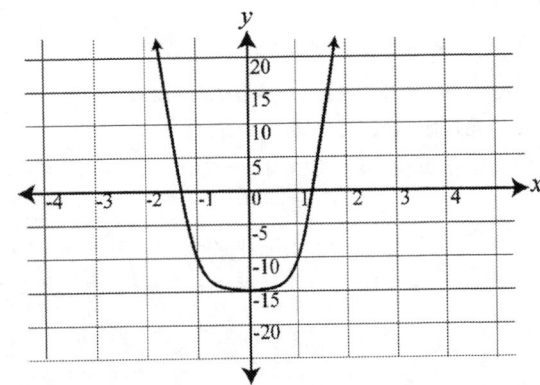

19. $f(x) = -7x^2 - 6$

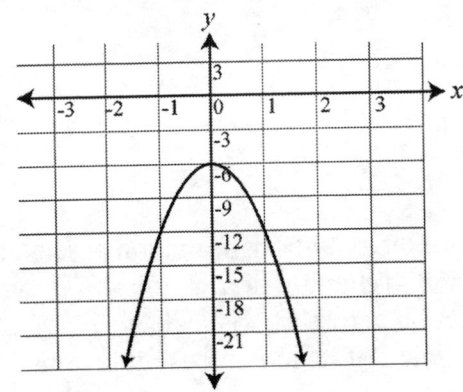

22. $f(x) = x^2 - 12x + 36$

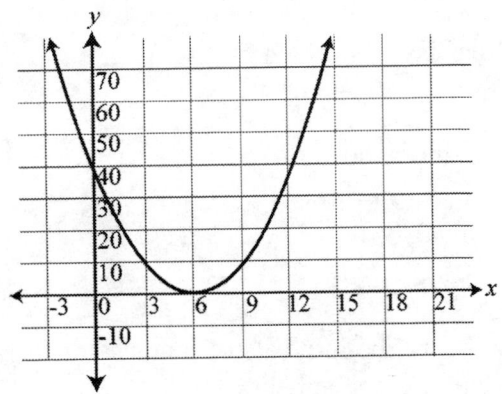

20. $f(x) = -x^3 + 24$

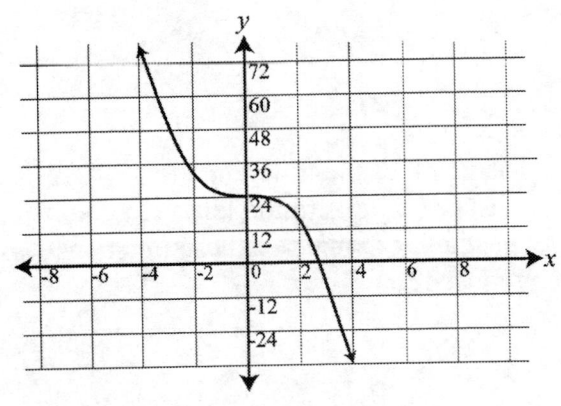

23. $f(x) = 2x^5 - 5x$

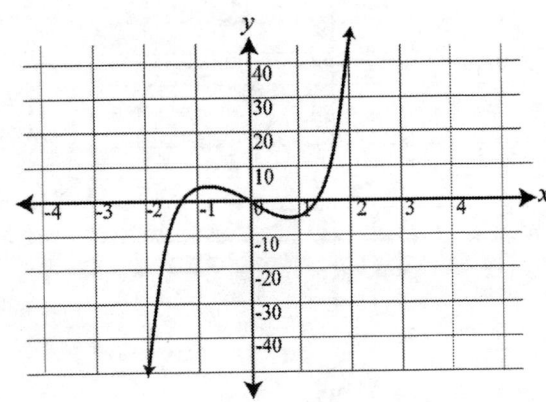

Chapter 8 Test

1. A polynomial function has a range of $y \geq 22$. Which of these must the function have?

   A. A relative maximum
   B. A relative minimum
   C. An absolute maximum
   D. An absolute minimum

2. What is the y-intercept of the graph of the function $f(x) = x^3 - 12x^2 - x + 12$?

   A. $(0, 12)$
   B. $(0, -1)$
   C. $(0, 1)$
   D. $(0, 12)$

3. What is the domain of the function $f(x) = -x^2 - 50$?

   A. $(-\infty, -50]$
   B. $(-\infty, \infty)$
   C. $[-50, \infty)$
   D. $[0, \infty)$

4. What are the intervals of increase of the function graphed below?

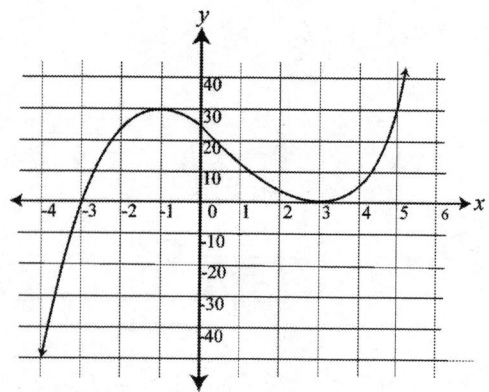

   A. $(-\infty, -3]$ and $[3, \infty)$
   B. $(-\infty, -1]$ and $[3, \infty)$
   C. $(-\infty, 1]$ and $[3, \infty)$
   D. $(-\infty, 3]$ and $[3, \infty)$

5. Suppose the only transformation applied to the graph of the function $f(x) = 6x^4$ was a reflection across y-axis. The result would be the graph of which of these functions?

   A. $f(x) = -6x^4 - 6$
   B. $f(x) = -6x^4$
   C. $f(x) = 6x^4$
   D. $f(x) = 6x^4 + 6$

6. If the graph of the function $f(x) = 10x^2$ were transformed by a horizontal stretch, the result could be the graph of which of these functions?

   A. $f(x) = 6x^2$
   B. $f(x) = 14x^2$
   C. $f(x) = 22x^2$
   D. $f(x) = 30x^2$

7. Suppose the only transformation applied to the graph of the function $f(x) = 27x^5$ was a reflection across x-axis. The result would be the graph of which of these functions?

   A. $f(x) = -27x^5 - 27$
   B. $f(x) = -27x^5$
   C. $f(x) = 27x^5$
   D. $f(x) = 27x^5 + 27$

8. The graph of the function $f(x) = 18x^7$ was translated 2 units to the left and 12 units up. The graph of which of these functions was produced?

   A. $f(x) = 18(x - 2)^7 - 12$
   B. $f(x) = 18(x - 2)^7 + 12$
   C. $f(x) = 18(x + 2)^7 - 12$
   D. $f(x) = 18(x + 2)^7 + 12$

9. Which of these functions has an absolute maximum?

   A. $f(x) = -26x^5$

   B. $f(x) = -7x^4$

   C. $f(x) = 9x^3$

   D. $f(x) = 11x^2$

10. A polynomial function has a zero of $x = 5$, which occurs two times, and a zero of $x = 8$, which occurs three times. Which of these statements is true?

    A. The multiplicity of $x = 5$ is odd, and the multiplicity of $x = 8$ is odd.

    B. The multiplicity of $x = 5$ is odd, and the multiplicity of $x = 8$ is even.

    C. The multiplicity of $x = 5$ is even, and the multiplicity of $x = 8$ is odd.

    D. The multiplicity of $x = 5$ is even, and the multiplicity of $x = 8$ is even.

11. Which of these statements accurately describes the function graphed below?

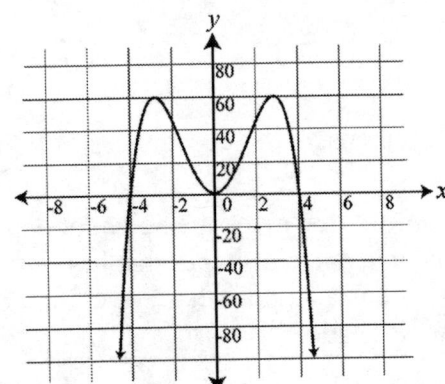

    A. It has an even degree and a negative lead coefficient.

    B. It has an even degree and a positive lead coefficient.

    C. It has an odd degree and a negative lead coefficient.

    D. It has an odd degree and a positive lead coefficient.

12. Which of these functions has a range of all real numbers?

    A. $f(x) = -18x^6 - 7$

    B. $f(x) = -14x^4 + 3$

    C. $f(x) = 12x^3 + 5$

    D. $f(x) = 16x^2 - 1$

13. If a polynomial function is odd, it is symmetric with respect to which of these?

    A. The x-axis

    B. The y-axis

    C. The origin

    D. The line $y = x$

14. The graph of which of these functions is symmetric with respect to the y-axis?

    A. $f(x) = x^2 + 20$

    B. $f(x) = x^2 + 20x$

    C. $f(x) = x^3 + 20$

    D. $f(x) = x^3 + 20x$

15. The graph of which of these functions is symmetric with respect to the origin?

    A. $f(x) = x^5 - 34$

    B. $f(x) = x^5 - 34x$

    C. $f(x) = x^6 - 34$

    D. $f(x) = x^6 - 34x$

16. The graph of the function shown below is symmetric with respect to which of these?

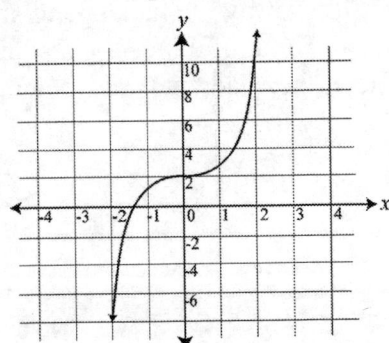

    A. The y-axis

    B. The origin

    C. The line $x = 2$

    D. The point $(0, 2)$

Chapter 9
Three-Dimensional Space

This chapter covers the following Georgia Performance Standards:

| MM3G | Geometry | MM3G3a, MM3G3b, MM3G3c |

9.1 Plotting the Point (x, y, z)

To determine the coordinates of a point on the xy-plane, follow a line perpendicular to the axes from the point to each axis and record the appropriate (x, y) value. The **quadrants** are labeled for your reference.

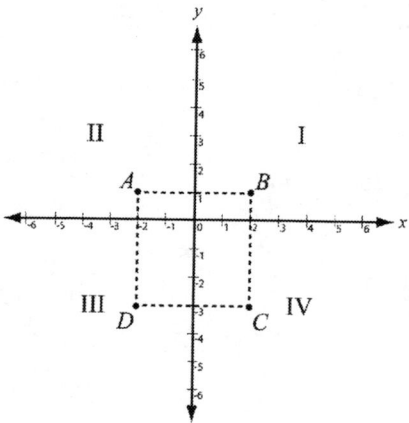

In this figure, the (x, y) coordinates of Point A are $(-2, 1)$. Likewise, the coordinates of Point B are $(2, 1)$, Point C $(2, -3)$ and Point D $(-2, -3)$.

In a 3D space, there is a third point, represented by the variable z, that gives the height or depth of the point. It is always a challenge to show three dimensions on a piece of paper that has only two dimensions. Think of the x, y-plane as a table top, and the z-axis as a pole that goes through the center. Here, the quadrants are labeled, and the first quadrant is shaded.

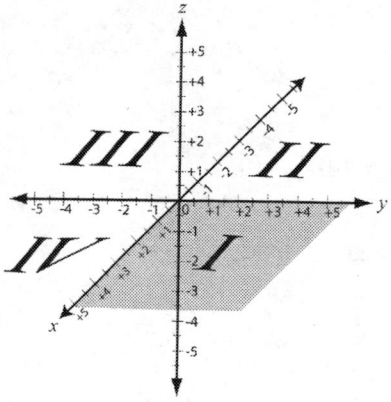

To determine the coordinates of a point on the xyz-plane, follow a line perpendicular to the axes from the point to each axis and record the appropriate (x, y, z) value.

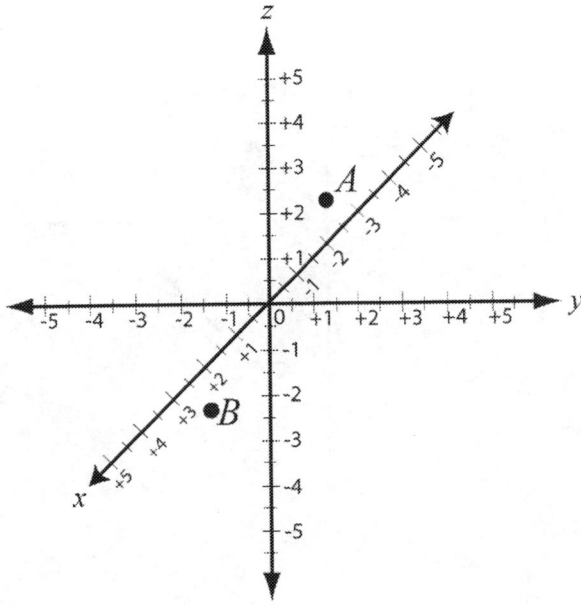

On the figure, Point A has coordinates $(1, 2, 3)$. Starting from the origin, move 1 unit to the forward ($x = 1$), then 2 units right ($y = 2$), then 3 units up ($z = 3$). Likewise, Point B is located at $(-1, -2, -3)$, or back 1, left 2, down 3.

Notice that each point in 3D space can be the vertex of a rectangular prism.

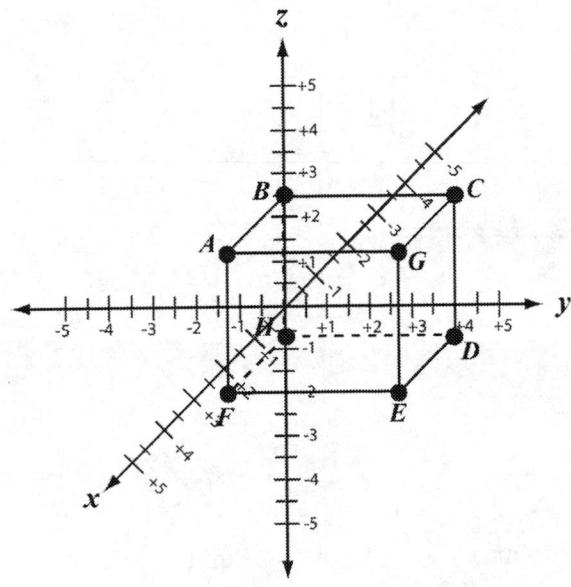

In the diagram below, identify the coordinates for:

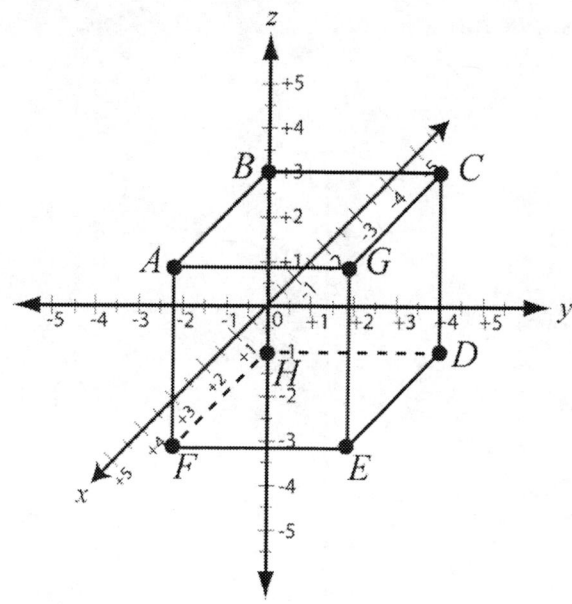

1. Point A	3. Point C	5. Point E	7. Point G
2. Point B	4. Point D	6. Point F	8. Point H

For the diagram, identify the (x, y, z) coordinates for:

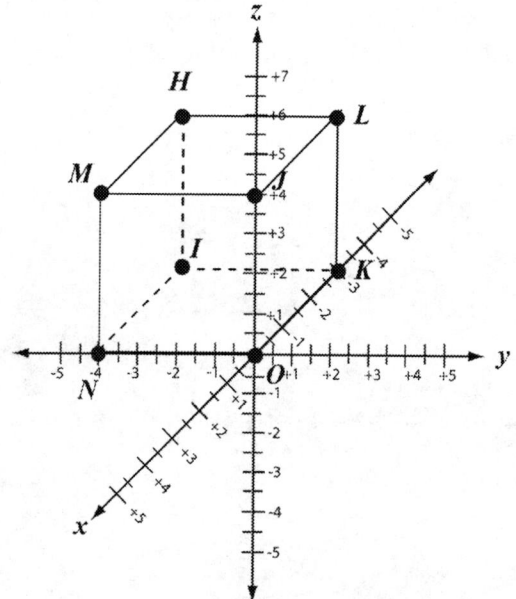

9. Point H	11. Point J	13. Point L	15. Point N
10. Point I	12. Point K	14. Point M	16. Point O

Write the ordered triple for each of the vertices of the following rectangular prism

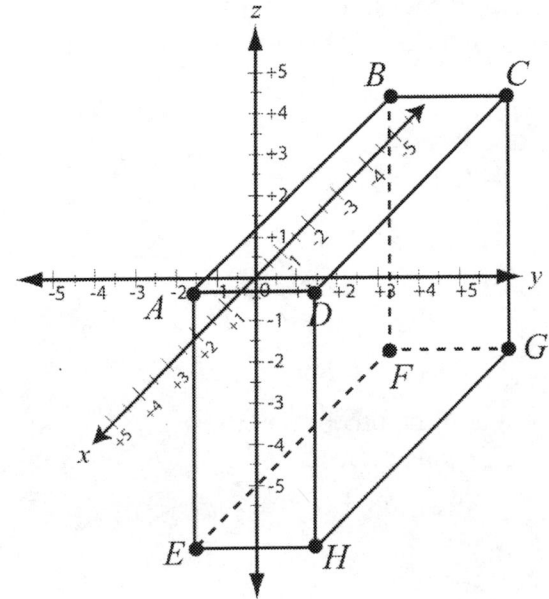

1. A (, ,)   5. E (, ,)

2. B (, ,)   6. F (, ,)

3. C (, ,)   7. G (, ,)

4. D (, ,)   8. H (, ,)

Plot and label the following points on the graph.

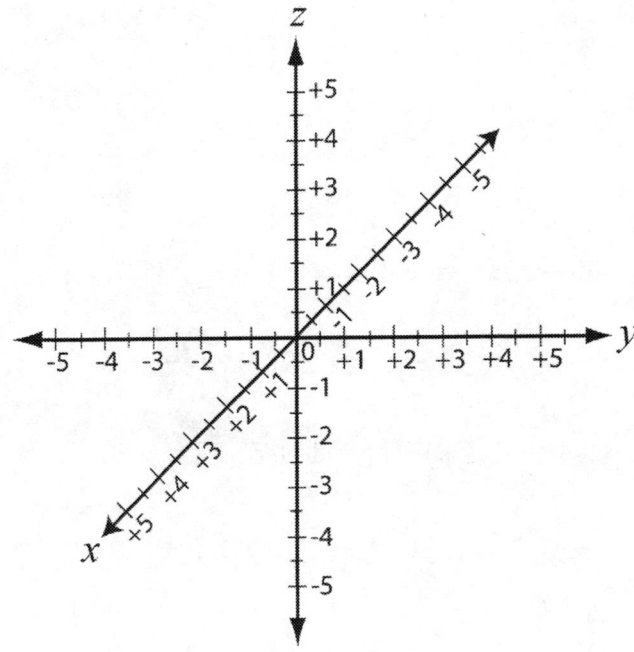

9. $A\,(2, 1, 4)$ 13. $E\,(2, 2, -4)$

10. $B\,(-1, 2, 5)$ 14. $F\,(-5, 3, -1)$

11. $C\,(-1, -5, 3)$ 15. $G\,(-4, -1, -1)$

12. $D\,(3, -2, 5)$ 16. $H\,(3, -3, -1)$

9.2   Applying the Distance Formula in 3-Space

The distance formula, $d = \sqrt{(x_2 - x_1)^2 + (y_2 - y_1)^2}$, is based on the Pythagorean Theorem and is used to calculate the distance between two points in a two dimensional space. Because there is an "extra" point in three-space, it follows that the **distance formula for three-space** would include that extra point.

$$d = \sqrt{(x_2 - x_1)^2 + (y_2 - y_1)^2 + (z_2 - z_1)^2}$$

Example 1:   Find the distance between the following two points in three-space:

$$A\,(1, 2, 3) \qquad \text{and} \qquad B\,(4, 6, 8)$$

Step 1:   Designate one of the points as $(x_1,\, y_1,\, z_1)$ and the other point as $(x_2,\, y_2,\, z_2)$. Substitute the values into the three-space distance formula.

$$d = \sqrt{(x_2 - x_1)^2 + (y_2 - y_1)^2 + (z_2 - z_1)^2}$$

$$d = \sqrt{(4 - 1)^2 + (6 - 2)^2 + (8 - 3)^2}$$

$$d = \sqrt{(3)^2 + (4)^2 + (5)^2}$$

$$d = \sqrt{9 + 16 + 25}$$

$$d = \sqrt{50}$$

$$d = 5\sqrt{2} \approx 7.1$$

Find the distance between Point A and Point B, and round your answer to the nearest tenth.

1. $A\,(-12, -9, -1)$ $B\,(1, 12, -3)$

2. $A\,(-10, -12, 7)$ $B\,(6, -5, -10)$

3. $A\,(5, 0, 6)$ $B\,(-6, 2, -11)$

4. $A\,(5, 12, -5)$ $B\,(-5, -5, 5)$

5. $A\,(-4, -9, 6)$ $B\,(-9, -4, -10)$

6. $A\,(-6, 3, 9)$ $B\,(-11, 4, 5)$

7. $A\,(-6, -7, 0)$ $B\,(-4, -2, -12)$

8. $A\,(-6, -6, 3)$ $B\,(-11, 12, 11)$

9.3   Understanding Equations of Planes and Spheres

Two points determine a line, and three non-collinear points determine a **plane**. Think of a plane like a sheet of paper, even though a sheet of paper has some thickness (though it is very, very small) and a plane has no thickness. The formula for the equation of a plane is

$$Ax + By + Cz = D$$

To sketch the graph of a plane, set two of the three terms equal to zero to find the point where the plane intersects the axes. Repeat for the other two variables.

Example 2:   Sketch the graph of the equation $4x + 8y - 16z = 32$.

Step 1:   To find the point where the plane intersects the x-axis, set $y = 0$ and $z = 0$.

$4x + 8(0) - 16(0) = 32$

$4x + 0 - 0 = 32$

$4x = 32$, so $x = 8$

This means that the plane intersects the x-axis at the point $(8, 0, 0)$

Step 2:   Next, let $x = 0$ and $z = 0$, so $8y = 32$, $y = \frac{32}{8}$, or $y = 4$.

This means that the plane intersects the y-axis at the point $(0, 4, 0)$.

Step 3:   To find where the plane intersects the z-axis, set x and $y = 0$, so $z = \dfrac{32}{-16}$ or -2. $(0, 0, -2)$

Step 4:   Plot those three points, connect them, then shade the plane that contains the three points.

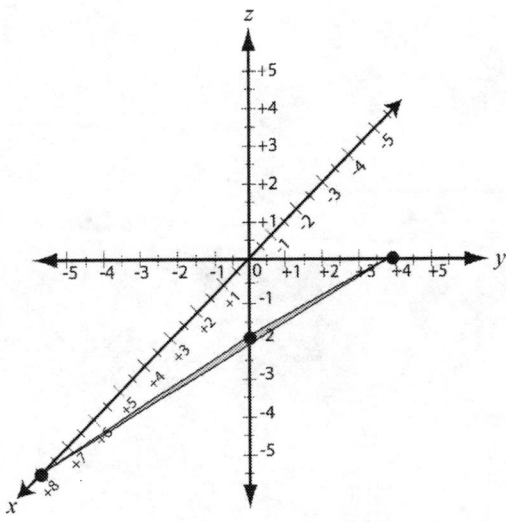

Sketch the graph of the following equations of planes.

1. $x + 5y - 2z = 10$

3. $4x - 9y - 18z = 36$

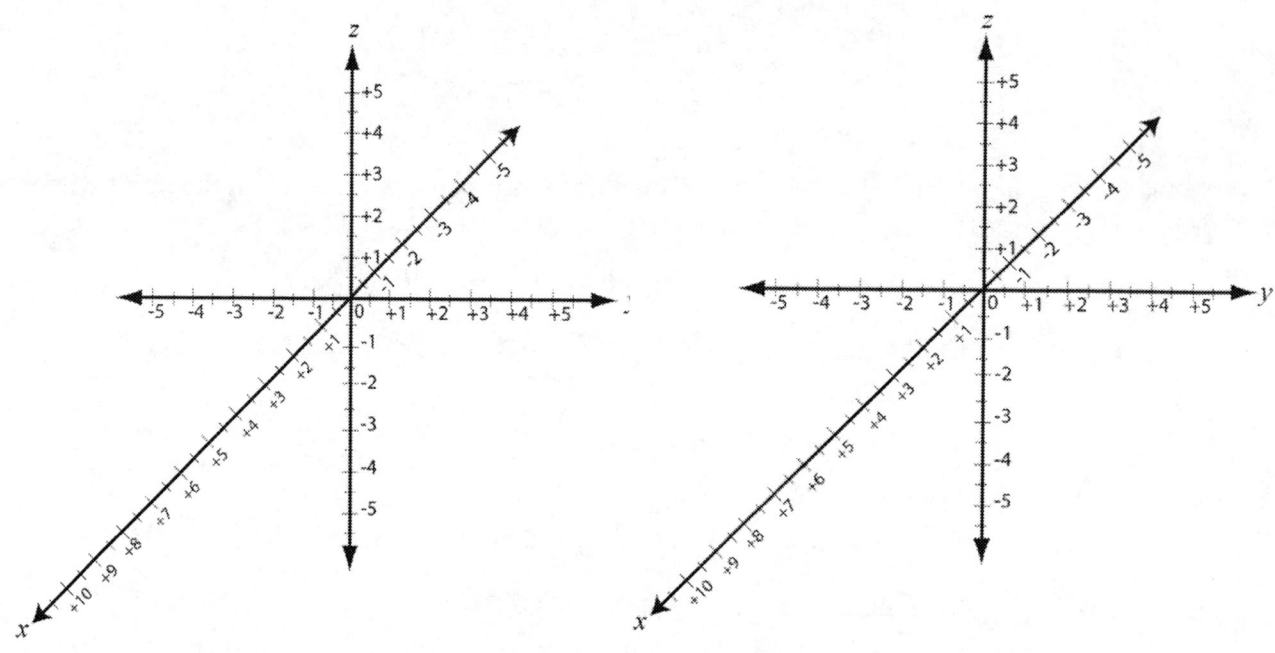

2. $6x + 3y + 12z = 12$

4. $3x - 15y + 15z = 30$

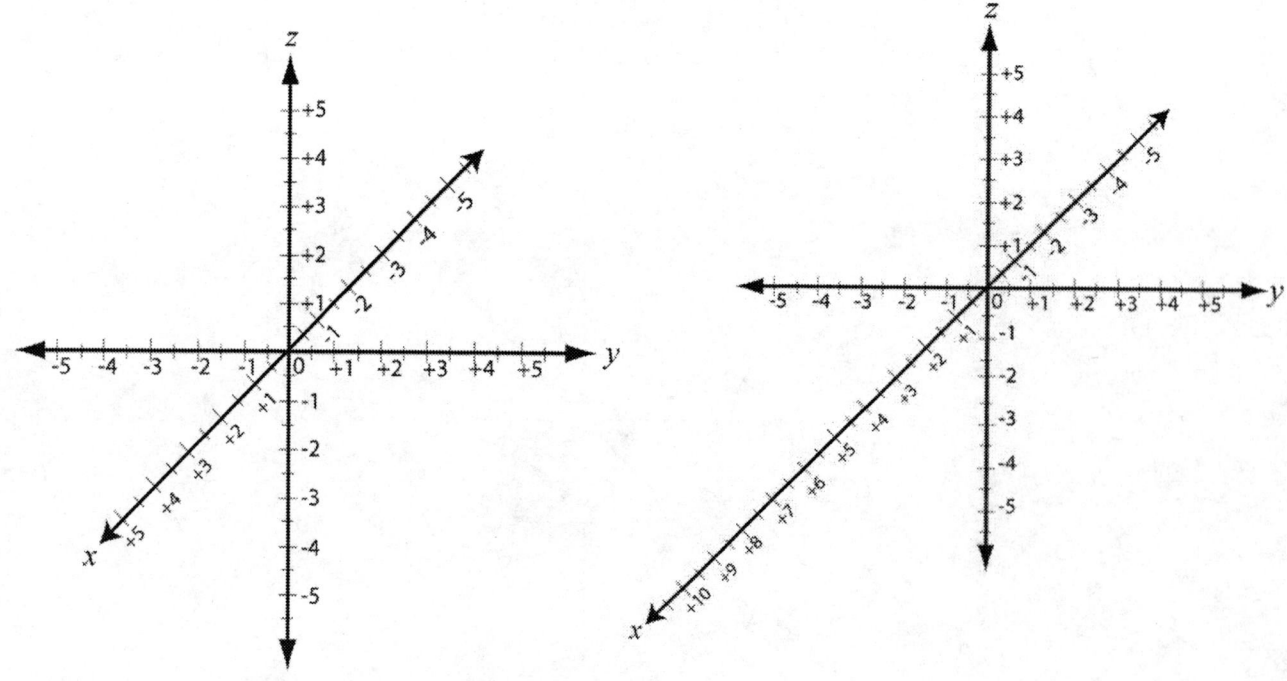

A **sphere** is the collection of all the points in 3-space that are the same distance from the center. A sphere is a ball, like a beachball, that has a radius, r, and a center, (x_0, y_0, z_0).

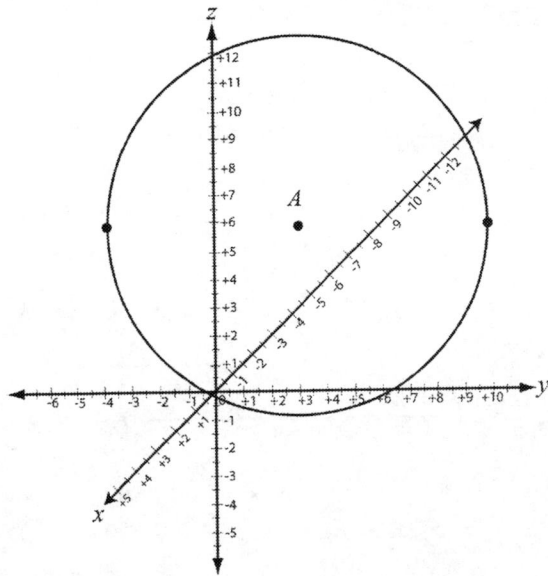

The equation of a sphere in standard form is

$$(x - x_0)^2 + (y - y_0)^2 + (z - z_0)^2 = r^2$$

Example 3: Given a sphere with a radius of 7 and center at $(-3, 1, 4)$, write the equation of the sphere in standard form.

Step 1: Plug the center point and radius into the equation of a sphere and simplify.

$$(x - x_0)^2 + (y - y_0)^2 + (z - z_0)^2 = r^2$$
$$(x - (-3))^2 + (y - 1)^2 + (z - 4)^2 = 7^2$$
$$(x + 3)^2 + (y - 1)^2 + (z - 4)^2 = 49$$

Write the equation of a sphere in standard form.

1. radius = 3, center = $(0, 2, 6)$

2. radius = 11, center = $(2, 7, 5)$

3. radius = 4, center = $(-9, 7, 1)$

4. radius = 8, center = $(1, -3, 6)$

5. radius = 2, center = $(3, -2, 1)$

6. radius = 1.6, center = $(9, -9, 4)$

7. radius = $\frac{2}{3}$, center = $\left(-\frac{1}{2}, \frac{1}{8}, \frac{1}{2}\right)$

8. radius = π, center = $(-\pi, 6\pi, 7\pi)$

Chapter 9 Review

1. Identify the following points on the diagram: Point A, Point B, Point C, Point D, Point E, Point F, Point G, and Point H.

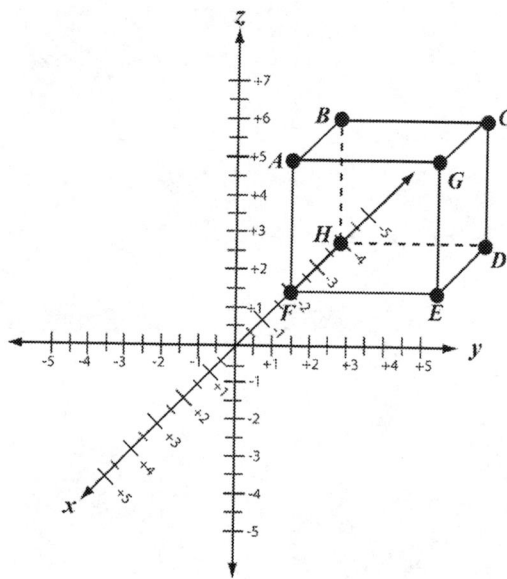

2. Find the distance between Point A and Point B, and round your answer to the nearest tenth.

$A\,(7, 7, -2)$ $\qquad$ $B\,(1, 2, -7)$

3. What are the three points where the plane, $-4x + 2y - 5z = 40$, crosses the x, y, and z axes?

4. Write the equation of a sphere in standard form.

radius $= 3$, center $= (0, 2, 6)$

Chapter 9 Test

Use the following graphic for questions 1–6.

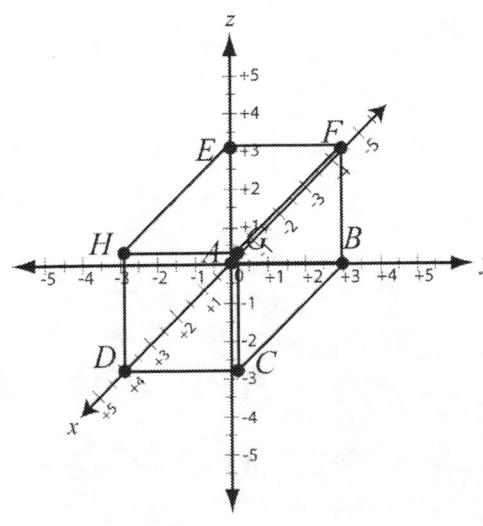

1. Point A is located at _____ .

   A. $(0, 0, 0)$

   B. $(0, 0, 1)$

   C. $(0, 1, 0)$

   D. $(1, 0, 1)$

2. Point H is located at _____ .

   A. $(3, 0, 4)$

   B. $(-3, 0, 4)$

   C. $(4, 0, 3)$

   D. $(3, 0, -4)$

3. Which point is located at $(0, 3, 3)$?

   A. A

   B. F

   C. E

   D. H

4. What is the measure of $\overline{AH}$?

   A. 5

   B. -5

   C. $\sqrt{7}$

   D. 8

5. Which solid is shown in the figure?

   A. rectangle

   B. cube

   C. rectangular prism

   D. sphere

6. What is the measure of $\overline{DF}$?

   A. 3.14

   B. 6.02

   C. 4.67

   D. 5.83

7. At what point does the plane, $5x - 3y + z = 150$, intersect the z-axis ?

   A. $(0, 0, 150)$

   B. $(30, 0, 0)$

   C. $(30, 50, 150)$

   D. $(150, 150, 150)$

8. What is the diameter of the following sphere?

   $(x - 6)^2 + (y + 4)^2 + (z - 8)^2 = 100$

   A. 10

   B. 100

   C. $(6, -4, 8)$

   D. 20

Chapter 10
Probability

This chapter covers the following Georgia Performance Standards:

MM3D	Data Analysis and Probability	MM3D1 MM3D2a, MM3D2b, MM3D2c

10.1   Creating a Probability Distribution

A **probability distribution** is a graph and table that displays the probability of each possible outcome of a random variable. These probabilities must all add up to equal 1.

Example 1:   Let X be the random variable that shows the number of heads that show after a person flips three quarters. Construct a probability distribution that represents the probability distribution of X.

Step 1:   Construct the table first to get all the information needed to construct the histogram. Create the first row with the variable X and number the columns 0 through 3 since that represents the number of heads that can possibly show in this situation. Label the second row as outcomes representing the number of possible ways that the corresponding number of heads can show. Finally, label the third row as $P(X)$ to represent the total number of flips considered and using that as the denominator of the probability fraction and using the corresponding positive outcome number as the numerator of the probability fraction.

X (**Number of heads**)	0	1	2	3
Outcomes	1	3	3	1
P(X)	$\frac{1}{8}$	$\frac{3}{8}$	$\frac{3}{8}$	$\frac{1}{8}$

Step 2:   Using the information from the table, construct a histogram that has $P(X)$ on the vertical axis and has X on the horizontal axis with the appropriate labels such as Probability and Number of Heads Shown.

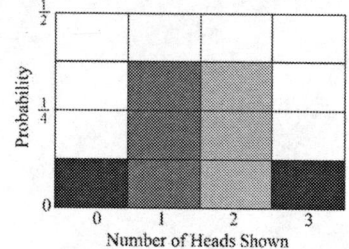

10.2 Interpreting Probability Distribution

Looking at the probability distributions, there are ways to interpret the findings and make statements about the probabilities.

Example 2: Recall the example in the previous section. What if you were asked to find the probability of having at least one head show when flipping the three quarters? This means that you can have $1, 2,$ or 3 heads show which includes the probability of all three outcomes.

Step 1: To figure this out, you must look at the table and find the corresponding probabilities and outcomes. Remember that the question asks to find the probability of having at least one head showing which means that the probability problem will be solved with the following equation:

$P(X \geq 1) = P(X = 1) + P(X = 2) + P(X = 3)$. Look back to the probability table.

X (**Number of heads**)	0	1	2	3
Outcomes	1	3	3	1
P(X)	$\frac{1}{8}$	$\frac{3}{8}$	$\frac{3}{8}$	$\frac{1}{8}$

Step 2: The probability of having at least one head show when flipping the quarter is

$P(X \geq 1) = P(X = 1) + P(X = 2) + P(X = 3) = \frac{3}{8} + \frac{3}{8} + \frac{1}{8} = \frac{7}{8} =$ $0.875 = 87.5\%$

Answer: If three quarters are flipped, the probability of having at least one head show is 87.5%.

1. A student takes a true-false quiz consisting of 5 questions. What is the probability that he guesses at least 80% of the questions correctly?

2. Four dimes are flipped. What is the probability of heads showing two times when flipping the four dimes?

10.3   Factorials

A **factorial** is a mathematical operation represented by an exclamation point (!). This mathematical operation is only performed on positive integers. When it is applied, the integer is multiplied by every integer less than it, including 1. By definition, zero factorial is equal to 1, and 1 factorial is equal to 1. Using mathematical symbols, $0! = 1$, and $1! = 1$. Other factorials are best learned by example.

Examples:
$$5! = 5 \times 4 \times 3 \times 2 \times 1 = 120$$
$$10! = 10 \times 9 \times 8 \times 7 \times 6 \times 5 \times 4 \times 3 \times 2 \times 1 = 3,628,800$$
$$6! = 6 \times 5 \times 4 \times 3 \times 2 \times 1 = 720$$
$$3! = 3 \times 2 \times 1 = 6$$
$$1! = 1$$
$$0! = 1$$

Factorials can also be used in mathematical operations such as multiplication and division. They are used to shorten the writing of some problems and are used to solve problems involving permutations, combinations, and the binomial theorem.

Examples:
$$\frac{6!}{3!} = \frac{6 \times 5 \times 4 \times \cancel{3} \times \cancel{2} \times \cancel{1}}{\cancel{3} \times \cancel{2} \times \cancel{1}} = 6 \times 5 \times 4 = 120$$

$$5! \times 4! = (5 \times 4 \times 3 \times 2 \times 1) \times (4 \times 3 \times 2 \times 1) = (120) \times (24) = 2880$$

Solve the following factorials.

1. $4!$

2. $7!$

3. $2!$

4. $8!$

5. $0!$

6. $11!$

7. $1!$

8. $\dfrac{5!}{3!}$

9. $3! \times 6!$

10. $7! \times 4!$

11. $\dfrac{8!}{5!}$

12. $0! \times 2!$

13. $\dfrac{10!}{0!}$

14. $4! \times 1!$

15. $\dfrac{8!}{3! \times 3!}$

16. $\dfrac{124!}{122!}$

17. $\dfrac{6!}{0! \times 4!}$

18. $\dfrac{12! \times 3!}{12!}$

10.4   The Combination Formula

The formula for combinations is used for counting problems. The formula is shown below.

$$_nC_k = \frac{n!}{k!\,(n-k)!}$$

The formula is read "n choose k."

Example 3:   Using the formula above, what is $_nC_k$ if $n = 5$ and $k = 2$?

Step 1:   Substitute that values for n and k into the formula.

$$_nC_k = \frac{n!}{k!\,(n-k)!} \to {_5C_2} = \frac{5!}{2!\,(5-2)!}$$

Step 2:   Solve.

$$_5C_2 = \frac{5!}{2!\,(5-2)!} = \frac{5!}{2!3!} = \frac{5 \times 4 \times 3 \times 2 \times 1}{(2 \times 1)(3 \times 2 \times 1)} = \frac{5 \times 4}{2 \times 1} = \frac{20}{2} = 10$$

Using the combination formula, find the following.

1. $_4C_0$

2. $_5C_1$

3. $_3C_0$

4. $_6C_1$

5. $_4C_1$

6. $_6C_2$

7. $_4C_2$

8. $_5C_3$

9. $_8C_3$

10. $_3C_1$

10.5   Creating a Binomial Distribution

A **binomial distribution** is a graph that shows the probabilities of the outcomes in a binomial experiment. A **binomial experiment** is a certain number of independent trials, denoted with the variable n, that have two possible outcomes. These outcomes are successes or failures in each trial. The success probability is the same for each trial. The formula for finding the number of exactly k successes for n trials is denoted as $P(k \text{ successes}) = {}_nC_k \, p^k (1-p)^{n-k}$.

Example 4:   A standard deck of cards contains 52 cards. Draw a histogram of the binomial distribution of drawing a card at random and that card being a diamond. Each card is replaced into the deck before the next one is drawn. The experiment is conducted a total of 7 times.

Step 1:   Start by computing the probability of each success probability using the equation, $P(k \text{ successes}) = {}_nC_k \, p^k (1-p)^{n-k}$. The probability of randomly drawing a diamond is 25%, so $p = 0.25$. The number of trials is 7, so $n = 7$. Since there are seven trials, there will have to be computations for each trial starting with $k = 0$ for there being no diamonds drawn in any of the seven trials. Then compute all the way to $k = 7$ which is where there are diamonds drawn each time in all 7 trials.

$$P(k = 0) = {}_7C_0 (0.25)^0 (0.75)^7 \approx 0.133$$
$$P(k = 1) = {}_7C_1 (0.25)^1 (0.75)^6 \approx 0.311$$
$$P(k = 2) = {}_7C_2 (0.25)^2 (0.75)^5 \approx 0.311$$
$$P(k = 3) = {}_7C_3 (0.25)^3 (0.75)^4 \approx 0.173$$
$$P(k = 4) = {}_7C_4 (0.25)^4 (0.75)^3 \approx 0.058$$
$$P(k = 5) = {}_7C_5 (0.25)^5 (0.75)^2 \approx 0.012$$
$$P(k = 6) = {}_7C_6 (0.25)^6 (0.75)^1 \approx 0.001$$
$$P(k = 7) = {}_7C_7 (0.25)^7 (0.75)^0 \approx 0.00006$$

Step 2:   Just like a probability distribution, create a binomial probability distribution by making a histogram of the probability results found in step 1. Set the vertical axis as Probability ranging from 0 to 0.4 in increments of 0.1. Set the horizontal axis as Number of Diamonds ranging from 0 to 7 in increments of 1 and leaving enough space to draw the bars.

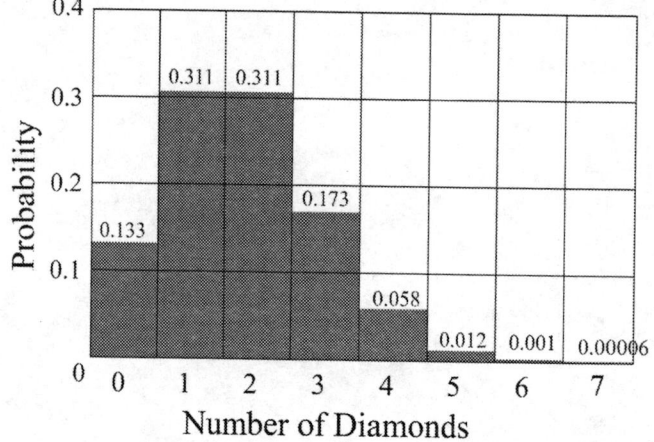

Example 5: A student is taking the Georgia Math III exam, which consists of 72 multiple choice questions with four choices per question. If a student guesses on every question, what is the probability that the student guesses only 36 correctly?

Step 1: Since there are four choices per question, the probability of guessing the correct answer is 0.25 so $p = 0.25$. There are 72 questions, so the number of trials is $n = 72$. The number of successes is 36, so $k = 36$. Plug these values into the equation, $P\left(k \text{ successes}\right) = {}_nC_k \, p^k \left(1 - p\right)^{n-k}$.

Step 2: $P\left(k \text{ successes}\right) = {}_nC_k \, p^k \left(1 - p\right)^{n-k}$
$P\left(k = 36\right) = {}_{72}C_{36} \left(0.25\right)^{36} \left(1 - \left(0.25\right)\right)^{72-36}$
$P\left(k = 36\right) = {}_{72}C_{36} \left(0.25\right)^{36} \left(0.75\right)^{36}$
$P\left(k = 36\right) = \left(4.4 \times 10^{20}\right)\left(2.1 \times 10^{-22}\right)\left(3.2 \times 10^{-5}\right)$
$P\left(k = 36\right) \approx 0.000003 \text{ or } 0.0003\%$

Answer: The probability of a student guessing 36 questions correctly on the exam is about 0.0003% or $\dfrac{3}{1,000,000}$.

1. According to the survey in the newspaper, about 10% of people in Atlanta are unemployed. Suppose you asked 5 random people in Atlanta if they were employed are not. On the space provided below, construct a histogram of the binomial distribution showing that exactly k of the people are unemployed. Label the bars with the probability values.

2. From the histogram created in question 1, what is the most likely outcome in the survey?

3. What is the probability that $k = 2$ from the histogram in question 1?

4. Looking at the histogram in question 1, is the histogram skewed or symmetric? This means either the histogram is not symmetric around a single line or the histogram is symmetric around a single line.

10.6   Calculating Binomial Probabilities with Graphing Calculators

Now that you know how to create a binomial distribution by hand, the graphing calculator can create a histogram based on information that is put into it.

Example 6:   According to a recent survey, 86% of the Braves fans believe that Bobby Cox should not have retired as the Braves' Manager. Suppose you go to a Braves baseball game and survey 7 random Braves' fans. Draw a histogram of the binomial distribution showing the probability that exactly k people agree with the statement. What is the most likely number of people in your survey that agree with the statement?

Step 1:   Enter the values for k. Go into the graphing calculator and press STAT and select the option Edit... under the EDIT tab. Under L1 put the numbers 0 through 7.

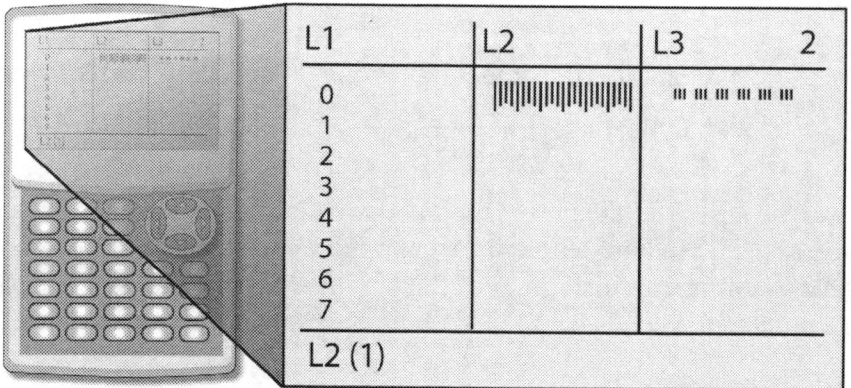

Step 2:   Find the values of $P(k)$. Press the 2nd button and select DISTR. Scroll down to find the binompdf(function, or press 0. With the calculator at the main screen there will be binompdf(and you must enter two things: the number of trials, 7, and the probability of success, 0.86. Press ENTER. This will give you all the $P(k)$ for each trial. The results must be stored into L2. Select STO⇒ and press 2nd, L2. Your calculator should look like this:

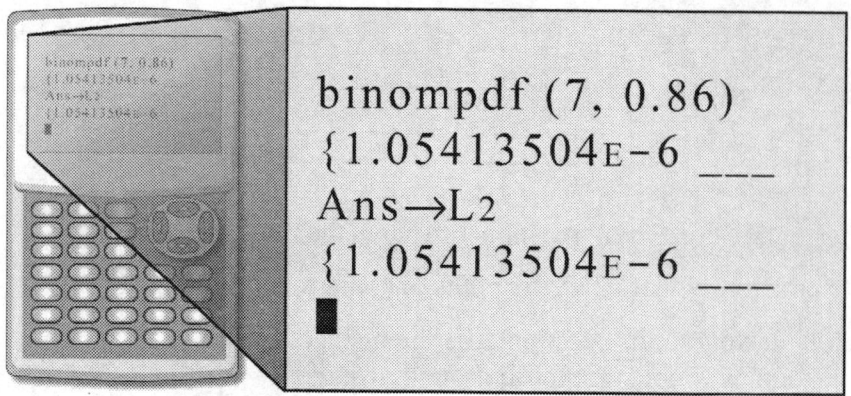

Step 3:   Set up the histogram. Press 2nd STAT PLOT. Turn Plot 1 to On by selecting the first Plot. In the same window, find the histogram drawing in the Type category. Set Xlist to L1 and Freq to L2. Select the WINDOW button and set the Xmin to 0 and Xmax to 9 so that it includes all values. Set the Ymin to 0 and Ymax to 0.5 since this will include all probabilities. Press GRAPH. Your window should look like this:

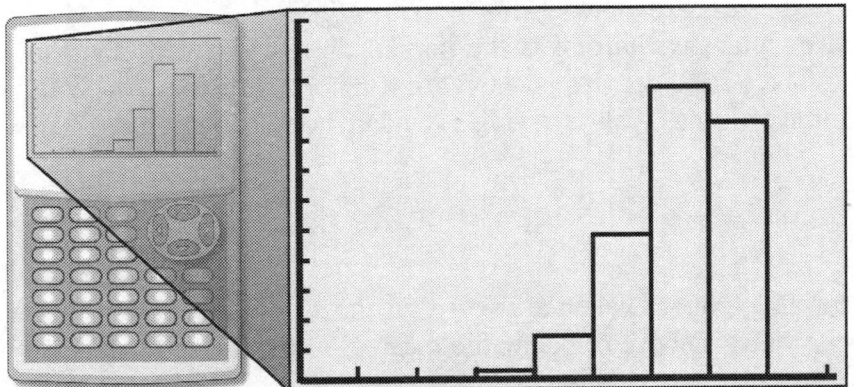

Step 4:   Find the most likely number of people who agree with the statement. To do this, press TRACE. Use the arrows to get to the tallest bar and record what is at the min= part.

Answer: The most likely number of people in a survey of 7 random people that agree that Bobby Cox should not have retired as the Braves' Manager is 6.

For the following questions, find the most likely number of successes using your graphing calculator given that a binomial experiment consists of n trials with a probability p of success in each trial with exactly k successes.

1. $n = 6, p = 0.14$

2. $n = 24, p = 0.70$

3. $n = 32, p = 0.49$

4. $n = 16, p = 0.32$

5. $n = 22, p = 0.66$

6. $n = 14, p = 0.11$

10.7   Normal Distribution

A **normal distribution** is a bell-shaped curve model that is called a **normal curve** which is symmetric about the mean. The normal distribution has a mean denoted by $\overline{x}$ and a standard deviation denoted by σ. **Standard deviation** is the measure of the variability (or spread) of the values in a data set. The total area under the normal curve is 1 and can have certain areas under the curve that is the standard deviation distance from the mean.

The **standard normal distribution** is the normal distribution with a mean of 0 and standard deviation of 1. The way to convert x-values from a normal distribution with a mean of $\overline{x}$ and a standard deviation of σ into z-values having a standard normal distribution is done by the following formula:

$$z = \frac{x - \overline{x}}{\sigma}$$

The z-value for an x-value is called a **z-score** for the x-value. It shows the number of standard deviations the x-value is above or below the mean. We convert to z-values to have one specific chart for any type of normal distribution.

To find the probability that z is less than or equal to some given value, use the standard normal table below.

Standard Normal Table									
z	.0	.1	.2	.3	.4	.5	.6	.7	.8	.9
−3	0.0013	0.0010	0.0007	0.0005	0.0003	0.0002	0.0002	0.0001	0.0001	0.0000+
−2	0.0228	0.0179	0.0139	0.0107	0.0082	0.0062	0.0047	0.0035	0.0026	0.0019
−1	0.1587	0.1357	0.1151	0.0968	0.0808	0.0668	0.0548	0.0446	0.0359	0.0287
−0	0.5000	0.4602	0.4207	0.3821	0.3446	0.3085	0.2743	0.2420	0.2119	0.1841
0	0.5000	0.5398	0.5793	0.6179	0.6554	0.6915	0.7257	0.7580	0.7881	0.8159
1	0.8413	0.8643	0.8849	0.9032	0.9192	0.9332	0.9452	0.9554	0.9641	0.9713
2	0.9772	0.9821	0.9861	0.9893	0.9918	0.9938	0.9953	0.9965	0.9974	0.9981
3	0.9987	0.9990	0.9993	0.9995	0.9997	0.9998	0.9998	0.9999	0.9999	1.0000−

The normal distribution curve and standard normal distribution curve are shown in the graphs below.

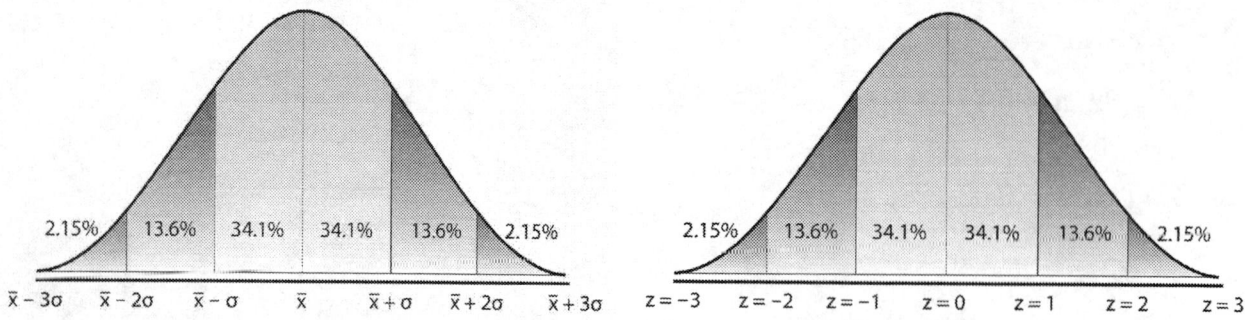

Note: The probability outside $\overline{x} \pm 3\sigma$ and $z = \pm 3$ is 0.15% for each end.

Use these graphs and table to help you solve problems contained in the next few sections.

10.8   Find a Normal Probability

Example 7:   A normal distribution has a mean of $\overline{x}$ and a standard deviation of σ. For a randomly selected x-value from the distribution, find $P(\overline{x} - 2\sigma \leq x \leq \overline{x} + \sigma)$.

Step 1:   Refer to the normal distribution curve on page 216. Find the corresponding percentages in each section inside the given interval.

Step 2:   The probability of selecting a random x-value from the distribution between $\overline{x} - 2\sigma$ and $\overline{x} - \sigma$ is 13.6% or 0.136. The probability of selecting a random x-value from the distribution between $\overline{x} - \sigma$ and $\overline{x}$ is 34.1% or 0.341. The probability of selecting a random x-value from the distribution between $\overline{x}$ and $\overline{x} + \sigma$ is 34.1% or 0.341.

Step 3:   Add the probability values together to find the probability within the whole interval. $P(\overline{x} - 2\sigma \leq x \leq \overline{x} + \sigma) = 0.136 + 0.341 + 0.341$

Answer: $P(\overline{x} - 2\sigma \leq x \leq \overline{x} + \sigma) = 0.818$ or 81.8%.

A normal distribution has a mean of $\overline{x}$ and a standard deviation of σ. For a randomly selected x-value, find the indicated probability. Give answer in percent form.

1. $P(\overline{x} - 2\sigma \leq x \leq \overline{x} + 2\sigma)$

2. $P(\overline{x} - 3\sigma \leq x \leq \overline{x} + \sigma)$

3. $P(x \leq \overline{x} + \sigma)$

4. $P(\overline{x} - 3\sigma \leq x)$

5. $P(\overline{x} - 3\sigma \leq x \leq \overline{x} + 3\sigma)$

6. $P(\overline{x} - \sigma \leq x)$

10.9   Interpreting Normally Distributed Data

There are ways of telling how much of a population will fall within a given interval when the mean, population size, and standard deviation are known. Figuring this out is very similar to predicting the probability of selecting a random value within a normally distributed population.

Example 8:   The heights of 150 males who are juniors in high school at a particular high school are normally distributed with a mean of 71 inches and a standard deviation of 2.5 inches. About how many of these high school juniors have heights from 66 inches to 73.5 inches?

   Step 1:   Find out how many standard deviations each end of the interval is from the mean by subtracting the mean from the value and dividing it by the standard deviation. $\left(\dfrac{66 - 71}{2.5} = \dfrac{-5}{2.5} = -2 \right)$ This means that the minimum value is 2 standard deviations below the mean and, according to the normal distribution curve, this includes 34.1% and 13.6% of the data. $\left(\dfrac{73.5 - 71}{2.5} = \dfrac{2.5}{2.5} = 1 \right)$ This means that the maximum value is 1 standard deviation above the mean and, according to the normal distribution curve, this includes 34.1% of the data.

   Step 2:   Find the total percent of the data that is included in the desired interval. $34.1\% + 13.6\% + 34.1\% = 81.8\%$

   Step 3:   Multiply the percent by the population size to get how many students fall within the interval. $0.818 \times 150 = 122.7$. Since there cannot be a decimal for people, it is safe to round the number to 123.

   Answer: About 81.8% or 123 of the males who are juniors at this particular high school have heights from 66 inches to 73.5 inches.

1. A survey shows that the number of words per minute a person can type is normally distributed with a mean of 50 wpm and a standard deviation of 15 wpm. According to the survey, what percent of people can type between 20 wpm and 95 wpm?

2. A cereal box company has boxes of cereal with the weight of each box being normally distributed with a mean of 15 ounces and a standard deviation of 1.2 ounces. If you were to randomly select 500 boxes of cereal from this company, about how many would you expect to weigh at least 16.2 ounces?

3. A restaurant sells drinks that are 22 ounces, but it has been shown that the drinks are actually normally distributed in volume with a mean of 21.8 ounces and a standard deviation of 0.8 ounces. Out of the 1000 drinks sold yesterday, how many do you expect to have been at most 20.2 ounces?

4. A survey of 800 people was conducted on how long they spend in a shopping center. The data was normally distributed with a mean of 48 minutes and a standard deviation of 8 minutes. About how many people from the survey do you expect to have spent 40 minutes to 72 minutes in the shopping center?

10.10   Z-Scores and the Standard Normal Table

Using the standard normal table and z-scores, you can find the probability of randomly picking a certain value out of a normally distributed set of data. The standard normal table and z-score formula are found on page 216.

Example 9:   The weight of a standard paper clip is normally distributed with a mean of 1.4 grams and a standard deviation of 0.3 grams. What is the probability of randomly selecting a paper clip and it having a weight of at most 2.09 grams?

Step 1:   Use the z-score formula to find the z-score corresponding to an x-value of 2.09.

$z = \dfrac{x - \overline{x}}{\sigma}$, where x is the given value (2.09), $\overline{x}$ is the given mean (1.4), and σ is the given standard deviation (0.3).

$$z = \frac{(2.09) - (1.4)}{0.3}$$
$$z = \frac{0.69}{0.3}$$
$$z = 2.3$$

Step 2:   Use the standard normal table on page 216 to find $P\left(x \leq 2.09\right) = P\left(z \leq 2.3\right)$. The table shows that $P\left(z \leq 2.3\right) = 0.9893$.

Answer: The probability that a randomly selected paper clip weighs less than or equal to 2.09 grams is 0.9893 or 98.93%.

1. What is the z-score for an x-value of 72 if all the values are normally distributed with a mean of 86 and a standard deviation of 10?

2. The number of hours a person sleeps every night is normally distributed with a mean of 8.1 hours and a standard deviation of 0.5 hours. What is the probability of randomly selecting a person and he has slept at most 6.9 hours?

3. A survey showed that the number of pages a person can read in an hour is normally distributed with a mean of 56 pages and a standard deviation of 2.4 pages. What is the probability that you randomly selected a person that could read at most 53 pages?

4. A certain neighborhood has the cost of painting a house normally distributed with a mean of $7,600$ and a standard deviation of 610. What is the probability that a randomly selected house was painted for at most $8,027$?

Chapter 10 Review

A normal distribution has a mean of $\overline{x}$ and a standard deviation of σ. Find the probability of a randomly selected x-value from the distribution within the given interval.

1. $P\left(x \leq \overline{x} - 3\sigma\right)$

2. $P\left(\overline{x} - 3\sigma \leq x \leq \overline{x} + 3\sigma\right)$

3. $P\left(\overline{x} - 2\sigma \leq x \leq \overline{x} + \sigma\right)$

4. $P\left(x \leq \overline{x} + \sigma\right)$

5. $P\left(\overline{x} - \sigma \leq x \leq \overline{x} + 2\sigma\right)$

6. $P\left(\overline{x} - \sigma \leq x\right)$

7. $P\left(\overline{x} + 2\sigma \leq x\right)$

8. $P\left(x \leq \overline{x} + 2\sigma\right)$

For the following questions, find the most likely number of successes using your graphing calculator given that a binomial experiment consists of n trials with a probability p of success in each trial with exactly k successes.

9. $n = 40, p = 0.76$

10. $n = 32, p = 0.11$

11. $n = 14, p = 0.84$

12. $n = 5, p = 0.89$

13. $n = 7, p = 0.81$

14. $n = 3, p = 0.06$

A normal distribution has a mean of 21 and a standard deviation of 3.5. Find the probability that a randomly selected x-value from the distribution is in the given interval.

15. Between 24.5 and 31.5

16. Between 10.5 and 21

17. Between 10.5 and 31.5

18. Between 14 and 24.5

A normal distribution has a mean of 422 and a standard deviation of 10. Use the standard normal table on page 216 to find the indicated probability for a randomly selected x-value from the distribution.

19. $P\left(x \leq 399\right)$

20. $P\left(x \leq 448\right)$

21. $P\left(x \leq 405\right)$

22. $P\left(x \leq 394\right)$

23. $P\left(x \leq 433\right)$

24. $P\left(x \leq 424\right)$

Four coins are flipped. Determine the probability of the following scenarios.

25. Exactly 2 heads shown.

26. At least 2 heads shown.

27. Only tails shown.

28. At least 3 tails shown.

Chapter 10 Test

1. An apple tree produces an average of 900 pounds of apples a year with a standard deviation of 120 pounds. If the data is normally distributed, what is the probability that a randomly selected tree from the distribution has produced up to $1,056$ pounds of apples?

   A. 0.0968
   B. -1.3
   C. 0.9032
   D. 1.3

2. In a recent survey, 73% of women use conditioner along with their shampoo. If you were to ask 15 random women if they use conditioner with their shampoo, how many would you expect to say that they do use conditioner with their shampoo based off this survey?

   A. 15
   B. 11
   C. 7
   D. 12

3. A normal distribution has a mean of 25 and a standard deviation of 5. What is the probability that a randomly selected x-value from the distribution is between 15 and 30?

   A. 81.8%
   B. 95.4%
   C. 68.2%
   D. 47.7%

4. When shooting a rubber band at a chain hanging from the ceiling, you have a 35% chance of hitting the chain. If you shot 15 rubber bands at the chain, how many rubber bands would you expect to hit the chain?

   A. 15
   B. 5
   C. 6
   D. 2

5. If $\overline{x} = 12$ and $\sigma = 3$, find $P\left(\overline{x} - 2\sigma \leq x \leq \overline{x}\right)$.

   A. 100%
   B. 68.2%
   C. 34.1%
   D. 47.7%

6. There are 300 pumpkins in a pumpkin patch. Their weights are normally distributed with a mean of 17.5 pounds and a standard deviation of 1 pound. How many pumpkins would you expect to weigh between 15.5 pounds and 18.5 pounds?

   A. 54
   B. 205
   C. 245
   D. 143

7. A company makes batteries that have a life span that is normally distributed. The average life span is 120 hours with a standard deviation of 8.7 hours. Any batteries that last longer than three standard deviations or shorter than three standard deviations from the mean are undesirable and will be discarded. What percent of the batteries would you expect to be discarded?

   A. 0.15%
   B. 0.015%
   C. 0.30%
   D. 30%

8. If there are 25 multiple choice questions with four possible answers per question, what would be the correct way to solve that you guessed 12 of the questions correctly?

   A. $P\left(k = 25\right) = {}_{25}C_{12}\left(0.25\right)^{12}\left(0.75\right)^{13}$
   B. $P\left(k = 12\right) = {}_{25}C_{12}\left(0.25\right)^{12}\left(0.75\right)^{13}$
   C. $P\left(k = 12\right) = {}_{12}C_{25}\left(0.25\right)^{12}\left(0.75\right)^{13}$
   D. $P\left(k = 25\right) = {}_{12}C_{25}\left(0.25\right)^{12}\left(0.75\right)^{13}$

Chapter 11
Data Interpretation

This chapter covers the following Georgia Performance Standards:

MM3D3	Data Analysis and Probability

11.1 Approximating Binomial Distributions Using Normal Distributions

A binomial distribution with n trials with the probability p of success on each trial can be approximated by a normal distribution if $np \geq 5$ and $n(1-p) \geq 5$. When these conditions are met the mean is now $\overline{x} = np$ and the standard deviation is now $\sigma = \sqrt{np(1-p)}$.

Example 1: According to a survey conducted by the National Resource Center, in 2008 about 72% of universities have a learning center. You are conducting a random survey of 500 universities. What is the probability of finding at most 350 universities that have a learning center?

Step 1: To approximate the binomial distribution with a normal distribution, the number of x universities in your survey that have a learning center has a binomial distribution with $n = 500$ and $p = 0.72$. To use the binomial distribution formula with this could be quite difficult, so we can approximate the answer using a normal distribution with the mean and standard deviation calculated by the formulas explained above.

$$\overline{x} = np = (500)(0.72) = 360$$

$$\sigma = \sqrt{np(1-p)} = \sqrt{(500)(0.72)(1-(0.72))} = \sqrt{100.8} \approx 10$$

Step 2: Now that the binomial distribution has been approximated by a normal distribution with a mean of 360 and a standard deviation of about 10, find how many standard deviations the desired number, 350, is from the mean.

$$\frac{350 - 360}{10} = \frac{-10}{10} = -1$$

This means that the desired number is about 1 standard deviation less than the mean.

Step 3: Use the normal distribution curve to find the approximate probability.

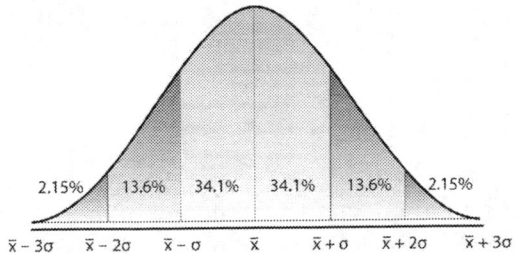

2.15% 13.6% 34.1% 34.1% 13.6% 2.15%

$\bar{x}-3\sigma$ $\bar{x}-2\sigma$ $\bar{x}-\sigma$ $\bar{x}$ $\bar{x}+\sigma$ $\bar{x}+2\sigma$ $\bar{x}+3\sigma$

Note: The probability outside $\bar{x} \pm 3\sigma$ is 0.15% for each end.

This means that $P(x \leq 350) \approx 0.0015 + 0.0215 + 0.136 = 0.159$.

Answer: The probability that at most 350 universities surveyed will have a learning center is about 0.159.

Answer the following questions about approximate probability with binomial distributions. Assume that all these binomial distributions can be approximated by a normal distribution. Round the standard deviations to the nearest whole number. Give the probabilities in decimal form.

1. According to a recent survey conducted by the National Health and Nutrition Examination Survey, about 30% of U.S. adults have high blood pressure. Suppose you conduct a random survey of 1,200 U.S. adults. What is the approximate probability that up to 392 U.S. adults will have high blood pressure?

2. According to the Manhattan Institute for Policy Research, in 1998 the U.S. high school graduation rate was 74%. If you were to ask a random group of 2100 people from the class of 1998 if they graduated from high school, what would be the approximate probability that up to 1554 people graduated high school?

3. According to the United States Department of Labor and the Bureau of Labor Statistics, in October 2010 9.6% of Americans are unemployed. If you were to conduct a random survey of 120 Americans, what is the approximate probability that up to 15 Americans are unemployed? Round the mean to the nearest whole number.

4. In a recent survey conducted by the United States Department of Labor and the Bureau of Labor statistics, in 2009 12.1% of Americans that are 16 to 24 years old make at or below minimum wage. Suppose you conduct a random survey of 3000 Americans that are 16 to 24 years old. What is the approximate probability that up to 327 of the people are unemployed?

11.2   Hypothesis Testing

When testing a hypothesis about a statistical measure for a population, there are certain steps to follow.

Step 1:   **State the hypothesis** that is being tested. This should include a statement about some statistical measure of a population.

Step 2:   **Collect the data** from a random sample of the population being tested, and compute the statistical measure of the sample.

Step 3:   **Assume the hypothesis** is true, and calculate the resulting probability, P, of obtaining the sample's statistical measure *or a more extreme* sample's statistical measure. If the probability is small, which is usually $P < 0.05$, then the hypothesis is rejected.

Example 2:   According to the Office for National Statistics and the General Lifestyle Survey, the United Kingdom has 34% of households had 2 people living in it in 2008. To test this finding, you survey 50 households in the United Kingdom and find that 15 of the households had 2 people living in it in 2008. Should you reject the Office for National Statistics' findings? Explain.

Step 1:   State the hypothesis: 34% of households in the United Kingdom have 2 people living in it in 2008.

Step 2:   Collect the data from a random sample of the population and calculate the statistical measure. In your survey 15 of the 50 households, or 30%, have 2 people living in it in 2008.

Step 3:   Assume that the hypothesis is true. Find the resulting probability that you could randomly select 15 *or fewer* households out of 50 that had 2 people living in it in 2008. The probability is:
$$P(x \le 15) = P(x = 0) + P(x = 1) + P(x = 2) + \ldots + P(x = 15),$$
where each term in the sum is a binomial probability with $n = 50$ and $p = 0.34$. The binomial distribution can be approximated with a normal distribution with the following mean and standard deviation:
$$\overline{x} = np = 50\,(0.34) = 17 \qquad \sigma = \sqrt{np\,(1-p)} = \sqrt{50\,(0.34)\,(0.66)} \approx 3.35$$
Using a z-score and the standard normal table on page 216, the approximate probability can be found.
$$P(x \le 15) \approx P\left(z \le \frac{15-17}{3.35}\right) \approx P(z \le -0.6) = 0.2743$$

Answer: So, if the survey is true that 34% of households in 2008 in the United Kingdom had 2 people living in the households, then there is about a 27% probability of finding 15 households or fewer that had 2 people living in it in 2008. With a probability this large, the hypothesis should *not* be rejected.

Answer the following questions by following the steps on the previous page. Use the space to show your work.

1. A recent survey conducted by the Center for Control Disease (CDC) has claimed that 24.3% of adult males in the United States have hypertension heart disease. Consider a random sample of 1,300 U.S. adult males. What is the probability that 359 or more will have hypertension? What is the probability that 293 or fewer will have hypertension? Round the mean and standard deviation to the nearest tenth.

2. According to a survey conducted by the CDC, 12.1% of people in the U.S. who are 75 years of age and older have had a stroke. Consider your random sample of 350 U.S. people who are 75 years of age or older and out of those 350 people, 32 people have had a stroke. Should you reject the claim made by the CDC? Explain.

11.3   Classifying Samples

Vocabulary

A **population** is a group of people or objects that there is a want of information from.

A **sample** is a subset of a population.

Self-selected Sample-members of a population volunteer to be in the sample.

Systematic Sample-a rule is used to select the sample from the population. For example, choosing every other person.

Convenience Sample-easy to collect sample such as picking people only from the front row of an audience.

Random Sample-each member of a population has an equal chance of being in the sample.

An **unbiased sample** gives a representation of the population that information is wanted from.

A **biased sample** is when the sample overrepresents or underrepresents the population that information is wanted from.

In the problems below, a survey is described. Identify each survey as convenience, self-selected, systematic, or random. Then, identify if the samples are biased or unbiased.

1. While on the phone with customer service, customers are asked if they would like to participate in a brief survey.

2. A survey is conducted at a grocery store where every male that walks into the store is included in a survey.

3. A magician is performing on stage and asks everyone in the front row to participate.

4. A person walks into Times Square and asks 250 random people to participate in a survey.

5. People in a room are given a number from 1 to 5, and the survey wants only the people who have the number 3 to participate.

6. A person walks door-to-door and asks if the family living there would like to participate in a voluntary survey.

7. Phone numbers are picked randomly from a phone book, then called, and a survey is conducted.

8. A farmer wants to know how many eggs his chickens are laying and counts the first 5 chickens he passes by.

11.4   Margin of Error

The **margin of error** describes how different the responses of a sample are compared to the responses of the whole population. The margin of error is approximated by the formula below.

$$\text{Margin of Error} = \pm\frac{1}{\sqrt{n}}$$

This means that the sample percentage that responded in a certain way is p, expressed as a decimal, and the population percentage that responded in the same way is likely to be between $p - \frac{1}{\sqrt{n}}$ and $p + \frac{1}{\sqrt{n}}$.

Example 3:   In a survey of $1,600$ people, 75% said that they go to the grocery store once a week. What is the margin of error for this survey? Determine an interval that is likely to contain the exact percent of all people who go to the grocery store once a week.

Step 1:   Use the margin of error formula.

$$\text{Margin of Error} = \pm\frac{1}{\sqrt{n}}$$

Step 2:   Substitute $1,600$ for n.

$$\text{Margin of Error} = \pm\frac{1}{\sqrt{1600}}$$
$$= \pm\frac{1}{40}$$
$$= \pm 0.025$$

This means that the margin of error for the survey is $\pm 2.5\%$.

Step 3:   Find the interval that is likely to contain the exact percent of all people who go to the grocery store once a week by subtracting and adding 2.5% to the percent of the people in the sample survey.
$$75\% - 2.5\% = 72.5\% \qquad 75\% + 2.5\% = 77.5\%$$

Answer: It is likely that the exact percent of all people who go to the grocery store once a week is between 72.5% and 77.5% given that the margin of error is 2.5%.

For problems 1 and 2, determine the margin of error and determine the interval that is likely to contain the exact percent of the population that has the same response.

1. A group of 900 high school students take a survey on how long they study each week. The survey showed that 30% of those high school students study 8 hours a week, on average.

2. In a survey of 500 households, 23% say that they have a dog as a pet.

There will be times when a margin of error is given and you are asked to find the sample size needed to obtain the given margin of error.

Example 4:   What sample size is needed to obtain a margin of error of $\pm 6.1\%$?

Step 1:   Work backwards from the margin of error formula. Be sure to change the margin of error to a decimal.

$$\text{Margin of Error} = \pm \frac{1}{\sqrt{n}}$$

$$\pm 0.061 = \pm \frac{1}{\sqrt{n}}$$

Since both sides of the equation have the $\pm$ symbol, they cancel each other out.

Step 2:   Square both sides to eliminate the square root.

$$(0.061)^2 = \left(\frac{1}{\sqrt{n}}\right)^2$$

$$0.003721 = \frac{1}{n}$$

Step 3:   Multiply both sides of the equation by n to eliminate the fraction.

$$0.003721\,(n) = \frac{1}{n}\,(n)$$

$$0.003721n = 1$$

$$n = \frac{1}{0.003721}$$

$$n = 268.744961$$

Answer: A sample size of 269 is needed in order to achieve a margin of error of $\pm 6.1\%$.

Determine the sample size required to achieve the given margin of error. Round the answer to the nearest whole number, if necessary.

1. $\pm 5.2\%$

2. $\pm 1.3\%$

3. $\pm 0.8\%$

4. $\pm 15.2\%$

5. $\pm 3.6\%$

6. $\pm 7.4\%$

7. $\pm 4.9\%$

8. $\pm 2.5\%$

9. $\pm 6.5\%$

10. $\pm 8.9\%$

11. $\pm 1.8\%$

12. $\pm 0.9\%$

13. $\pm 3.9\%$

14. $\pm 14\%$

15. $\pm 29\%$

16. $\pm 69\%$

17. $\pm 10\%$

18. $\pm 3.1\%$

11.5 Experimental and Observational Studies

There are two groups that are typically involved in a study. The first group is called the **experimental group** which has the test or study done to them. The other group is called the **control group** which does not have the test or study done to them. These two groups are as close in relation to each other as possible so that there is a way to figure out the effects of the tests or studies on the experimental groups. An **experimental study** occurs when the person who is conducting the tests or studies assigns individuals to a control group or an experimental group. An **observational study** occurs when the assigning of the individuals to the control group or the experimental group is out of the control of the person conducting the tests or studies.

Example 5: You want to study the effects of regular exercise on a person's blood pressure. You measure the blood pressure of a person at rest and again after they do jumping jacks for 5 minutes. The control group is the students who are not on an athletic team. The experimental group is the students who are on an athletic team. Is this an observational study or an experimental study? Explain.

Step 1: Recall the definitions of observational studies and experimental studies. Were the groups assigned by the observer? No.

Step 2: The groups were not controlled by the observer, but rather than by the choices the students made before the study to be on an athletic team or not.

Answer: This is an observational study because the groups were not assigned by the observer and were outside the control of the observer. The students were able to *sort themselves* into the two groups because of previously made decisions on whether or not to be on an athletic team.

Example 6: You want to do a study on the effects of headache medicine and see if the medicine is as effective as it claims to be or if it is more about the people thinking that they are getting headache medicine to relieve the pain. The experimental group is the people who have a headache and get the headache medicine. The control group is the people who have a headache and get the placebo (sugar) pill. Is this an observational study or an experimental study? Explain.

Step 1: Recall the definitions of observational studies and experimental studies. Were the groups assigned by the observer? Yes.

Step 2: The groups were controlled by the observer because you picked which ones got the medicine and which ones got the placebo pill.

Answer: This is an experimental study because the groups were assigned by the observer. The people who got the medicine were placed in the experimental group, and the people who got the placebo pill were placed in the control group.

Look at the following scenarios and describe them as either an observational study or an experimental study.

1. Scientists want to study the risk of developing lung cancer. The control group is non-smokers, and the experimental group is smokers.

2. A teacher wants to know if studying for a test really does help improve the chances of the student scoring 80% and above. The control group is the students who did not study for the test, and the experimental group is the students who did study for the test.

3. A company wants to measure the effects fertilizer has on plants. The control group is the plants that receive only sunlight and water. The experimental group is the plants that receive sunlight, water, and fertilizer.

4. It is rumored that a water bottling company sells water that may cause people to develop kidney stones. To study this, scientists give this type of water to hamsters and call this the experimental group. The control group is the group of hamsters that receive tap water.

5. You are sitting at a mall and want to know if, during the summer, people tend to walk around in flip flops. The control group is the people who walk around with something other than flip flops, and the experimental group is the people who walk around in flip flops.

6. A deodorant brand claims that they will not leave stains on shirts when applied. To test this claim, you assign 60 random males to use the deodorant and another 60 random males to not use deodorant. Then, you check the shirts of all of the random males for stains.

Chapter 11 Review

1. A recent study showed that 10% of all births in the U.S. result from a teen mom. You conduct a random survey of 600 new mothers. What is the approximate probability that at most 53 of those new mothers will be teenagers?

2. A survey showed that 14% of all Catholic people in the U.S. attend church every weekend. You survey 1,650 random Catholic people in the U.S. What is the approximate probability that at most 259 of the U.S. Catholics surveyed attend church every weekend?

3. You hear that a local pizza restaurant claims that 72% of their customers prefer pepperoni pizza. You want to test this finding. So you survey 72 of the customers in the restaurant and find that 41 of the customers prefer pepperoni pizza. Should you reject the claim made by the pizza restaurant? Explain.

4. At the last community meeting, the chairman claims that 10% of the community are not mowing their lawns enough which is supposed to be once a week. You survey 500 random people of the community and find that 61 of the people do not mow their lawn once a week. Should you challenge the claim made by the chairman by rejecting his claim? Explain.

5. A flight attendant wants to know if the passengers on the whole airplane are comfortable. She asks only the people in first class so that she can get a quicker response. What kind of sample is this, and is this a biased or unbiased sample?

6. A grocery store puts on their receipt that people can participate in a brief survey about the customer service and enter a chance to win a shopping gift card. What type of sample is it for the people who respond, and is this a biased or unbiased sample?

7. In a survey of 959 people, 12% said that they have been in a car wreck within the past year. What is the margin of error, and what is the interval that is most likely to contain the exact percent of all people who have been in a car wreck within the past year?

8. In a survey of 224 people, 9% said that they drink their coffee black. What is the margin of error, and what is the interval that is most likely to contain the exact percent of all people who drink their coffee black?

9. Scientists want to study the ability for monkeys to learn certain skills. They place 2 groups of monkeys in separate rooms. One room contains a video that shows how to place the blocks in order, and the other room contains just the blocks. Is this an observational study or an experimental study?

10. You want to know what percent of people get their bags checked at the airport. You observe the people when they are checking in for their flight and count how many check their bags. Are you conducting an experimental study or observational study?

Chapter 11 Test

1. You are told that the interval that is most likely to contain all the people who walk their dog every day is between 13.5% and 22.1%. What is the margin of error, and what sample size is needed to obtain that margin of error?

   A. ±4.3%; 32
   B. ±17.8%; 541
   C. ±4.3%; 541
   D. ±17.8%; 32

2. A recent survey showed that 40% of people like country music. You conduct a random survey of 10 people. What is the approximate probability that at most 3 people like country music?

   A. 0.4167
   B. 0.3446
   C. 41.67%
   D. Cannot be approximated because $np < 5$ and $n(1-p) > 5$.

3. You are having people stand in a line and numbering them $1, 2, 1, 2, 1....$ You are only interested in surveying the people who are numbered 1. What kind of sample are you taking?

   A. Self-selected Sample
   B. Systematic Sample
   C. Random Sample
   D. Convenience Sample

4. In a recent survey of 160 people, 31% said that they spend at least $50 a day on everyday expenses. What is the margin of error for this survey?

   A. about 7.91%
   B. about 0.079%
   C. about 0.625%
   D. about 14.20%

5. What kind of study is conducted where you have the ability to control what is placed in the experimental group and what is placed in the control group?

   A. Systematic Study
   B. Experimental Study
   C. Observational Study
   D. Population Study

6. A survey showed that 78% of people over 50 have had some sort of surgery in their life. If you randomly surveyed 700 people over 50 and found that 530 have had some sort of surgery in their life. Should you reject the claim made by the survey?

   A. No, because $P(z \leq -1.5) = 0.0668$ which is greater than 0.05.
   B. No, because $P(z \leq 10.96) = 0.0668$ which is greater than 0.05.
   C. Yes, because $P(z \leq -1.5) = 0.0668$ which is greater than 0.05.
   D. Yes, because $P(x \leq 530) = 0.0668$ which is less than 0.05.

7. A survey that is conducted where the person conducting the survey has no control over who is in the experimental group and who is in the control group is what type of study?

   A. Systematic Study
   B. Experimental Study
   C. Observational Study
   D. Population Study

8. You go to a subway station and pick out 50 random people to ask them if they use the subway as their main mode of transportation. What type of sample are you taking?

   A. Self-selected Sample
   B. Systematic Sample
   C. Random Sample
   D. Convenience Sample

Practice Test

Part 1

1. What is the equation of a circle that has a center at $(-4, 4)$ and a diameter of 6?

   A. $(x + 4)^2 + (y - 4)^2 = 9$
   B. $(x - 4)^2 + (y - 4)^2 = 9$
   C. $(x - 4)^2 + (y + 4)^2 = 81$
   D. $(x + 4)^2 + (y + 4)^2 = 81$

   MM3G1a

2. What is the equation of a circle that has a center at $(2, -9)$ and a radius of 7?

   A. $(x - 2)^2 + (y + 9)^2 = 49$
   B. $(x + 2)^2 + (y - 9)^2 = 49$
   C. $(x - 2)^2 + (y + 9)^2 = 7$
   D. $(x + 2)^2 + (y - 9)^2 = 7$

   MM3G1a

3. What is the equation of a circle that has a center 4 units below the origin and a radius of 1?

   A. $(x + 4)^2 + y^2 = 2$
   B. $x^2 + (y + 4)^2 = 1$
   C. $x^2 + (y - 4)^2 = 1$
   D. $(x - 4)^2 + y^2 = 2$

   MM3G1a

4. A circle has a radius of $2\sqrt{5}$ units with its center at the origin. What is the equation of a line that is tangent to this circle at the point $(-2, -4)$?

   A. $f(x) = 2x - 5$
   B. $f(x) = \frac{1}{2}x + 5$
   C. $f(x) = -\frac{1}{2}x - 5$
   D. $f(x) = -2x + 5$

   MM3G1c

5. A circle has a radius of $4\sqrt{13}$ units with its center at $(-2, 3)$. What is the equation of a line that is tangent to this circle at the point $(6, -9)$?

   A. $f(x) = -\frac{2}{3}x + 13$
   B. $f(x) = \frac{2}{3}x - 13$
   C. $f(x) = -\frac{2}{3}x - 13$
   D. $f(x) = \frac{2}{3}x + 13$

   MM3G1c

6. A circle has a radius of 25 units with its center at $(-21, -5)$. What is the equation of a line that is tangent to this circle at the point $(-45, -12)$?

   A. $f(x) = -\frac{24}{7}x + 166\frac{2}{7}$
   B. $f(x) = \frac{24}{7}x - 166\frac{2}{7}$
   C. $f(x) = \frac{24}{7}x + 166\frac{2}{7}$
   D. $f(x) = -\frac{24}{7}x - 166\frac{2}{7}$

   MM3G1c

7. How many times do the two circles that have equations $(x - 10)^2 + y^2 = 16$ and $(x - 7)^2 + y^2 = 16$ intersect, if any?

   A. The two circles intersect at one point.
   B. The two circles intersect at two points.
   C. The two circles intersect at infinitely many points.
   D. The two circles do not intersect.

   MM3G1e

8. If the equation to Circle A is $x^2 + y^2 = 36$ and the equation to Circle B is $4(0.5x)^2 + y^2 = 36$, how many points do the two circles intersect, if any?

   A. 0
   B. 1
   C. 2
   D. ∞

   MM3G1e

9. What does the graph of a circle with the equation $(x + 7)^2 + (y - 7)^2 = 49$ look like?

A.

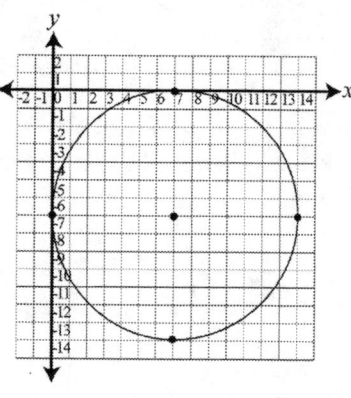

B.

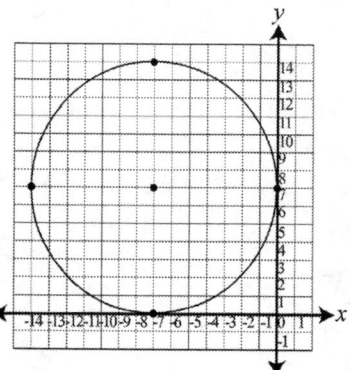

C.

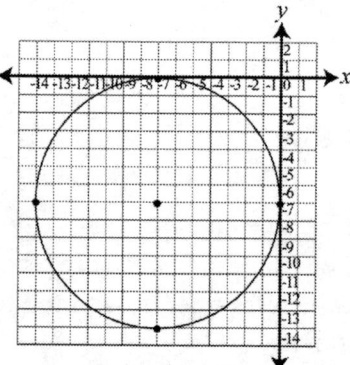

D.

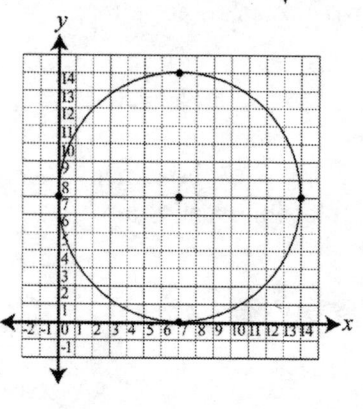

10. What does the graph of a circle with the equation $(x + 2)^2 + y^2 = 20$ look like?

A.

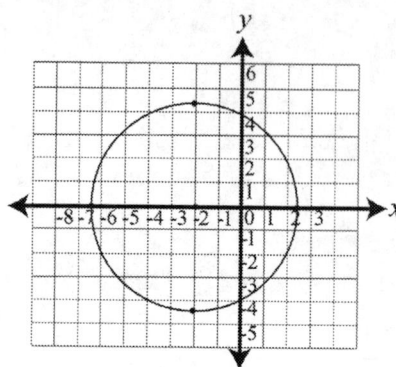

B.

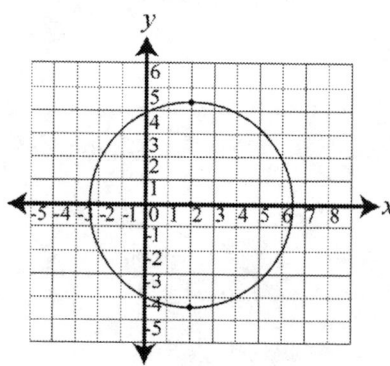

C.

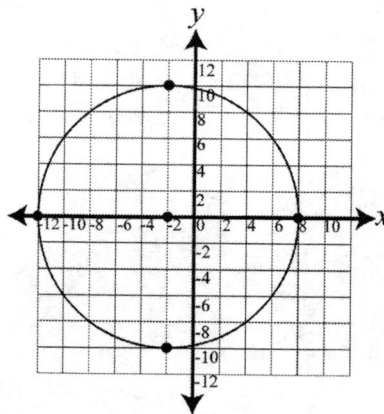

D.

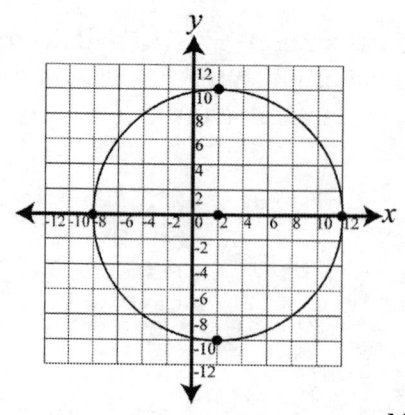

MM3G1b

MM3G1b

234

11. There are two points in the 3-space graph. Point A is located at $(-7, 2, 0)$ and Point B is located at $(4, 9, 10)$. What is the distance between the two points?

A. $3\sqrt{30}$

B. $2\sqrt{7}$

C. $\sqrt{158}$

D. 270

MM3G3b

12. What is the distance between two vertices of a rectangular prism if the two points are plotted at $(6, 9, 0)$ and $(-5, 1, -3)$?

A. 194

B. $\sqrt{22}$

C. $\sqrt{194}$

D. 2

MM3G3b

13. What is the equation of a sphere whose center is located at $(2, -5, 2)$ and the diameter measures $\sqrt{12}$ units.

A. $(x-2)^2 + (y+5)^2 + (z-2)^2 = 12$

B. $(x-2)^2 + (y+5)^2 + (z-2)^2 = 6$

C. $(x-2)^2 + (y+5)^2 + (z-2)^2 = 3$

D. $(x+2)^2 + (y-5)^2 + (z+2)^2 = 12$

MM3G3c

14. An equation of a plane is $-5x - 2y - 10z = -17$. Where does the plane cross the y-axis?

A. $\left(0, \dfrac{17}{2}, 0\right)$

B. $\left(0, 0, \dfrac{17}{10}\right)$

C. $\left(\dfrac{17}{5}, 0, 0\right)$

D. cannot be determined

MM3G3c

15. Find the zeros of the parabola by completing the square of the equation $y - 2x^2 + 16x = 5$.

A. $y = \dfrac{8 \pm 3\sqrt{6}}{2}$

B. $x = \dfrac{4 \pm 3\sqrt{6}}{2}$

C. $x = \dfrac{8 \pm 3\sqrt{6}}{2}$

D. $y = \dfrac{4 \pm 3\sqrt{6}}{2}$

MM3G2a

16. Classify the conic that has an equation $4x^2 - 24x - 9y^2 = 36$

A. Circle

B. Hyperbola

C. Ellipse

D. Parabola

MM3G2a

17. Convert the equation of the hyperbola to standard form.
$9x^2 - 90x - 16y^2 - 64y = -17$

A. $\dfrac{(x+5)^2}{16} - \dfrac{(y+2)^2}{9} = 1$

B. $\dfrac{(x-5)^2}{16} - \dfrac{(y+2)^2}{9} = 1$

C. $\dfrac{(x-5)^2}{9} - \dfrac{(y-2)^2}{16} = 1$

D. $\dfrac{(x+5)^2}{9} - \dfrac{(y+2)^2}{16} = 1$

MM3G2a

18. What is the equation of an ellipse that has a vertex at $(7, 0)$, a focus at $(4, 0)$, and its center is at the origin?

   A. $\dfrac{x^2}{49} + \dfrac{y^2}{33} = 1$

   B. $\dfrac{x^2}{7} + \dfrac{y^2}{33} = 1$

   C. $\dfrac{x^2}{33} + \dfrac{y^2}{49} = 1$

   D. $\dfrac{x^2}{33} + \dfrac{y^2}{7} = 1$

19. What is the equation of a circle that has a center at $(2, 5)$ and its circumference is 50π units?

   A. $(x + 2)^2 + (y + 5)^2 = 25$
   B. $(x + 2)^2 + (y + 5)^2 = 625$
   C. $(x - 2)^2 + (y - 5)^2 = 25$
   D. $(x - 2)^2 + (y - 5)^2 = 625$

20. A hyperbola has an equation of $\dfrac{x^2}{7} - \dfrac{y^2}{2} = 1$. What are the coordinates for the foci?

   A. $(3, 0)$ and $(-3, 0)$
   B. $(0, 3)$ and $(0, -3)$
   C. $(9, 0)$ and $(-9, 0)$
   D. $(0, 9)$ and $(0, -9)$

21. What is the length of a diagonal in a rectangular prism that has endpoints with coordinates of $(4, 3, -4)$ and $(8, 10, 9)$?

   A. 24
   B. 21.77
   C. 15.3
   D. 234

22. A hyperbola has an equation of $\dfrac{x^2}{676} - \dfrac{y^2}{169} = 1$. What is the equation for the asymptotes?

   A. $y = \pm 2x$

   B. $y = \pm \dfrac{676}{169} x$

   C. $y = \pm \dfrac{1}{2} x$

   D. $y = \pm \dfrac{169}{676} x$

23. What is the length of a diagonal in a rectangular prism that has endpoints with coordinates of $(-2, -5, 7)$ and $(-1, -2, 5)$?

   A. 3.74
   B. 14
   C. 6
   D. 2.45

24. If the equation of a circle is $(x + 7)^2 + (y + 4)^2 = 9$ and the equation of a line is $y = -4$, what are the point(s) of intersection, if any?

   A. $(-4, -4)\ (4, -4)$
   B. $(-4, -4)\ (-10, -4)$
   C. $(4, -4)\ (-10, -4)$
   D. no intersection

25. If the equation of a circle is $(x - 4)^2 + (y - 8)^2 = 4$ and the equation of a line is $x = 6$, what are the point(s) of intersection, if any?

   A. $(2, 8)$ and $(6, 8)$
   B. $(6, 8)$ and $(4, 8)$
   C. $(6, 8)$
   D. no intersection

26. Suppose that a quantity's doubling time is 50 years. How many times its original value will it be after 150 years?

    A. 2 times its original value
    B. 3 times its original value
    C. 6 times its original value
    D. 8 times its original value

MM3A2g

27. What is $\left(36^{\frac{1}{4}}\right)^{\frac{3}{5}}$?

    A. $36^{\frac{3}{20}}$

    B. $36^{\frac{7}{20}}$

    C. $36^{\frac{13}{20}}$

    D. $36^{\frac{17}{20}}$

MM3A2b

28. If the point $\left(-2, \frac{1}{5}\right)$ satisfies a function, which of these points satisfies its inverse?

    A. $\left(-2, \frac{1}{5}\right)$

    B. $\left(-\frac{1}{5}, 2\right)$

    C. $\left(\frac{1}{5}, -2\right)$

    D. $\left(2, -\frac{1}{5}\right)$

MM3A2c

29. If a logarithmic function increases throughout its domain, which of these statements must be true?

    A. Its base must be less than 0.
    B. Its base must be between 0 and 1.
    C. Its base must be 1.
    D. Its base must be greater than 1.

MM3A2e

30. Which of these transformations was applied to the graph of the function $f(x) = \log x$ to produce the graph shown below?

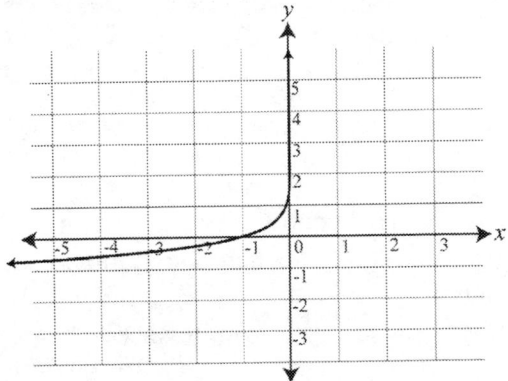

    A. Only a reflection across the x-axis
    B. Only a reflection across the y-axis
    C. Both a reflection across the x-axis and a reflection across the y-axis
    D. Neither a reflection across the x-axis and a reflection across the y-axis

MM3A2f

31. Which of these equations is in standard form?

    A. $x^3 + 3x^2 = 40 + 18x$
    B. $x^3 + 3x^2 = 18x + 40$
    C. $x^3 + 3x^2 - 40 - 18x = 0$
    D. $x^3 + 3x^2 - 18x - 40 = 0$

MM3A3d

32. If $6^x = 442$, which of these equations must be true?

    A. $\dfrac{x}{\ln 6} = 442$

    B. $\dfrac{x}{\ln 6} = \ln 442$

    C. $x \cdot \ln 6 = 442$

    D. $x \cdot \ln 6 = \ln 442$

MM3A3b

33. What is the solution to the equation
$\log_3 (x^2 + 4x + 4) = 2$?

A. $x = -5$ and -1
B. $x = -5$ and 1
C. $x = 5$ and -1
D. $x = 5$ and 1

MM3A3b

34. What is the x-coordinate of the point on the coordinate grid where the graphs of the functions $f(x) = \ln(x+2) + \ln(x+9)$ and $g(x) = 2\ln(x+5)$ intersect?

A. -43
B. -7
C. 7
D. 43

MM3A3b

35. What is the solution to the inequality $x^3 + x^2 - 44x - 84 < 0$ written in interval notation?

A. $(-6, -2) \cup (7, \infty)$
B. $[-6, -2] \cup [7, \infty)$
C. $(-\infty, -6) \cup (-2, 7)$
D. $(-\infty, -6] \cup [-2, 7]$

MM3A3c

36. If both sides of the inequality $x \cdot \ln\left(\frac{1}{9}\right) < 295$ are divided by $\ln\left(\frac{1}{9}\right)$ to solve for x, should the direction of the inequality sign be changed so that it becomes a greater-than sign?

A. No, because $\ln\left(\frac{1}{9}\right)$ is negative.

B. No, because $\ln\left(\frac{1}{9}\right)$ is positive.

C. Yes, because $\ln\left(\frac{1}{9}\right)$ is negative.

D. Yes, because $\ln\left(\frac{1}{9}\right)$ is positive.

MM3A3d

Part 2

37. What is the solution to the inequality $\log_{25}(x + 10) > 1$ in interval notation?

    A. $(-10, \infty)$

    B. $[-10, \infty)$

    C. $(15, \infty)$

    D. $[15, \infty)$

    MM3A3c

38. The graph of the function $f(x) = \log(x + 40) - 2$ is shown below.

    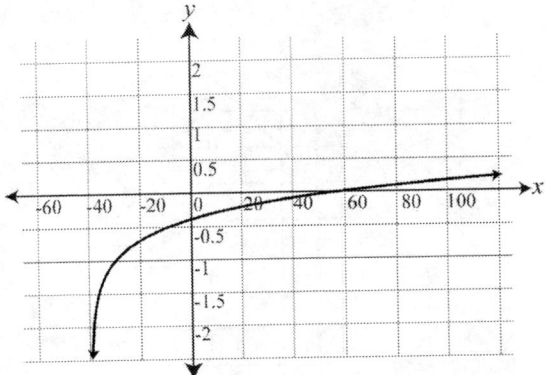

    What is the solution to the inequality $\log(x + 40) - 2 \geq 0$ in interval notation?

    A. $(-40, \infty)$

    B. $[-40, \infty)$

    C. $(60, \infty)$

    D. $[60, \infty)$

    MM3A3c

39. At how many points does the graph of the function $f(x) = x^3 + 12x^2 + 48x + 64$ intersect the x-axis?

    A. 0

    B. 1

    C. 2

    D. 3

    MM3A1d

40. If graph of the function $f(x) = 18x^3$ were transformed by a vertical stretch, the result could be the graph of which of these functions?

    A. $f(x) = 3x^3$

    B. $f(x) = 9x^3$

    C. $f(x) = 15x^3$

    D. $f(x) = 21x^3$

    MM3A1a

41. What are the zeros of the function graphed below?

    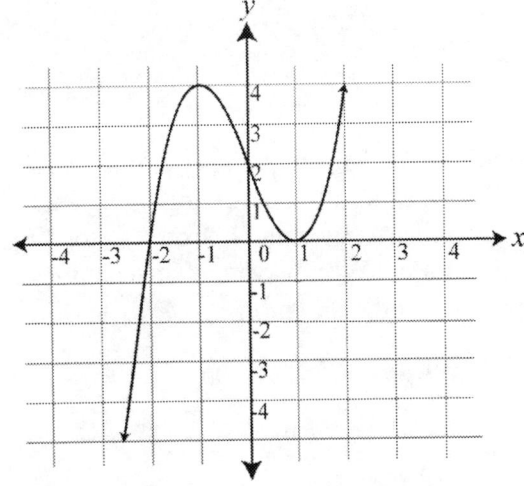

    A. $x = -2$, with an even multiplicity, and $x = 1$, with an odd multiplicity

    B. $x = -2$, with an odd multiplicity, and $x = 1$, with an even multiplicity

    C. $x = -1$, with an even multiplicity, and $x = 2$, with an odd multiplicity

    D. $x = -1$, with an odd multiplicity, and $x = 2$, with an even multiplicity

    MM3A1b

42. The graph of the function shown below is symmetric with respect to which of these?

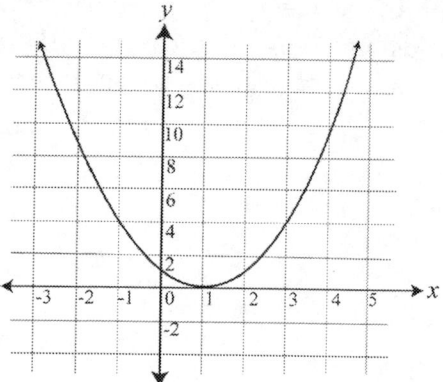

A. The y-axis

B. The origin

C. The line $x = 1$

D. The point $(0, 1)$

MM3A1c

43. Jack fetched seven pails, p, and six buckets, b, of water totaling 32 gallons. Jill fetched four pails, p, and nine buckets, b, totaling 35 gallons. Which system of equations represents the situation above?

A. $7p + 6b = 32; 4p + 9b = 35$

B. $7p + 4b = 32; 6p + 9b = 35$

C. $11p + 15b = 67; p + b = 26$

D. $11p + 15b = 67; p + b = 2$

MM3A5c

44. Dante can buy 15 markers and 9 pens at the school store for \$13.47 or he can buy 4 markers and 14 pens for \$6.84. How much is the price of a marker?

A. \$0.28

B. \$0.93

C. \$0.73

D. \$1.30

MM3A6b

45. Which graph correctly depicts the following system of inequalities?

$$y \geq 3x + 2$$
$$2x - 2y < -2$$

A.

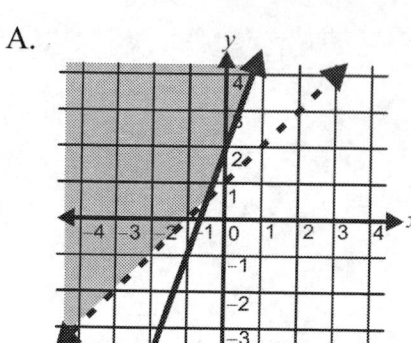

B.

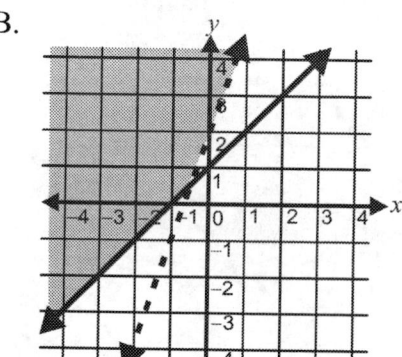

C.

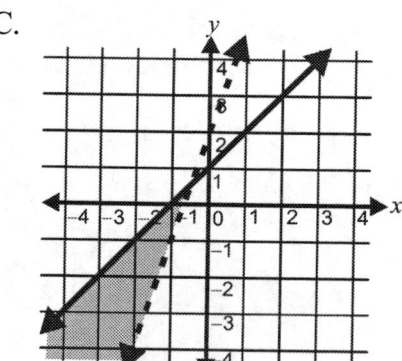

D.

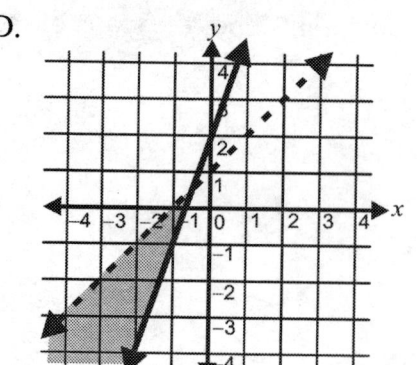

MM3A6a

46. Deserie plays a game where she spins a spinner numbered $24, 81, 68, 88, 89, 4, 62, 29$. If she lands on an even number she wins points. If she lands on an odd number she loses points. She gets to spin 10 times per round. The first round she landed on an even number 6 times and an odd number 4 times with a total of 64 points. The next round she landed on an even number 3 times and an odd number 7 times with a total of -8 points. How many points are won by landing on an even number?

A. 16

B. 15.73

C. -10.13

D. -3

MM3A6b

47. Draw a network represented by the matrix below.

$$\begin{array}{c|cccc} & A & B & C & D \\ \hline A & 0 & 0 & 1 & 1 \\ B & 0 & 0 & 1 & 1 \\ C & 1 & 1 & 0 & 1 \\ D & 1 & 1 & 1 & 0 \end{array}$$

A.

B.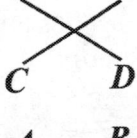

C. A B C D

D.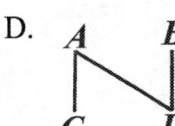

MM3A7b

48. $-2 \begin{bmatrix} -5 & -2 \\ 9 & -7 \\ 3 & 6 \end{bmatrix} + \begin{bmatrix} -1 & -2 \\ -8 & 7 \\ 9 & -4 \end{bmatrix} =$

A. $\begin{bmatrix} -6 & -4 \\ 1 & 0 \\ 12 & 2 \end{bmatrix}$

B. $\begin{bmatrix} -12 & -8 \\ 1 & 0 \\ 12 & 2 \end{bmatrix}$

C. $\begin{bmatrix} 9 & 2 \\ -26 & 21 \\ 3 & -16 \end{bmatrix}$

D. $\begin{bmatrix} 9 & -26 & 3 \\ 2 & 21 & -16 \end{bmatrix}$

MM3A4a

49. When taking a test that consists of 23 questions that have 5 answer choices each and only one correct answer choice, what is the approximate probability that you guessed 4 correctly?

A. 0.1633

B. 0.1940

C. 0.2042

D. 0.2000

MM3D1

50. There are 5 cards on the table where a magician has told you to pick a card and place it back without revealing your choice. The magician says that he will be able to find your card 4 out of 5 times. What is the probability that his claim is true?

A. 0.4096

B. 0.3277

C. 0.0064

D. 0.0512

MM3D1

51. Given that a binomial distribution has a probability of $p = 0.46$ and $n = 17$, what is the most likely number of successes that will occur?

A. 7

B. 6

C. 8

D. 5

MM3D1

52. Suppose you roll a die 8 times. What is the probability of a 3 or a 4 showing up at least 2 times?

A. 0.195

B. 0.844

C. 0.333

D. 0.167

MM3D1

53. If a data set that is normally distributed has a mean of $1,214$ and a standard deviation of 5, what is the interval that will contain 2.30% of the data set?

A. greater than $1,224$

B. between $1,224$ and $1,229$

C. greater than $1,229$

D. less than $1,199$

MM3D2a

54. What is the interval about the mean of a normally distributed data set that will include 84.1% of the data that has a mean of 36 and a standard deviation of 7?

A. from 29 to 57

B. greater than 57

C. less than 29

D. greater than 29

MM3D2a

55. What is the interval about the mean of a normally distributed data set that will include 68.2% of the data that has a mean of 439 and a standard deviation of 42?

A. from 397 to 439

B. from 397 to 481

C. from 439 to 481

D. cannot be determined

MM3D2a

56. What is the interval about the mean of a normally distributed data set that will include 99.7% of the data that has a mean of 74 and a standard deviation of 7.1?

A. from 52.7 to 95.3

B. from 52.7 to 74

C. all of the data

D. cannot be determined

MM3D2a

57. What percent of a normally distributed data set would fall between 9.1 and 20.2 if the mean is 12.8 and the standard deviation is 3.7?

A. 68.2%

B. 47.7%

C. 95.4%

D. 81.8%

MM3D2a

58. A normally distributed data set has a mean of 29 and a standard deviation of 3.7. What is the probability that a randomly selected value from the data set is between 21.6 and 32.7?

A. 0.682

B. 0.818

C. 818%

D. 68.2%

MM3D2b

59. What is the probability that a randomly selected value from a normal distribution that has a mean of 117 and a standard deviation of 21.5 is at most 172.9?

A. 2.6
B. 0.0047
C. −2.6
D. 0.9953

MM3D2b

60. The times that 120 high school students can run a 40-yard dash is normally distributed with a mean of 5.2 seconds and a standard deviation of 0.2 seconds. How many of those high school students do you expect to have run the 40-yard dash between 4.6 seconds and 5 seconds?

A. 3
B. 19
C. 16
D. cannot be determined

MM3D2c

61. What is the probability that a randomly selected value from a normal distribution that has a mean of 1234 and a standard deviation of 12.34 will be at least 1250.042?

A. 0.9680
B. 0.9032
C. 0.0968
D. cannot be determined

MM3D2b

62. The heights of 45 high school juniors are normally distributed with a mean of 66 inches with a standard deviation of 3 inches. How many of those juniors would you expect to be between 62.7 inches and 72.3 inches?

A. 6
B. 38
C. 44
D. 37

MM3D2c

63. The expected time batteries that are made at a company are supposed to last is normally distributed with a mean of 120 hours and a standard deviation of 8.7 hours. If the company produced 12 million batteries last year, how many would you expect to have lasted between 137.4 hours and 146.1 hours?

A. 258,000
B. 11,742,000
C. 2,850,000
D. 1,174,000

MM3D2c

64. What is the probability that a randomly selected value from a normal distribution that has a mean of 98 and a standard deviation of 13.2 will be between 101.96 and 141.56?

A. 0.3816
B. 0.9995
C. 0.8849
D. cannot be determined

MM3D2b

65. The amount of soda in a can at a soda can production plant has their volumes normally distributed with a mean of 11.2 ounces and a standard deviation of 0.8 ounces. If the factory produced 600 cans in one hour, how many of those cans do you expect to be at least 12 ounces?

A. 16
B. 95
C. 300
D. 94

MM3D2c

66. A student's average in a Math 3 course is normally distributed with a mean of 87 and a standard deviation of 4. Out of 300 students taking the Math 3 course, how many would you expect to have an average of at most 82?

A. 271
B. 102
C. 41
D. 29

MM3D2c

67. A recent study showed that 15% of people use a manual lawn mower rather than a riding lawn mower. You survey 50 households in your neighborhood and find that 7 of the households use manual lawn mowers. Should you reject the claim made by the study?

A. Yes, because $P(z \leq -0.2) \approx 0.4207$ which is greater than 0.05.

B. No, because $P(z \leq -0.2) \approx 0.4207$ which is greater than 0.05.

C. Yes, because $P(z \leq 0.2) \approx 0.5793$ which is greater than 0.05.

D. No, because $P(z \leq 0.2) \approx 0.5793$ which is greater than 0.05.

MM3D3

68. A recent study showed that 9.6% of Americans are unemployed. What is the approximate probability that out of a sample of 350 randomly selected Americans that up to 22 of the Americans are unemployed?

A. 0.0015

B. 0.0215

C. 0.0230

D. cannot be determined

MM3D3

69. A waitress is wanting to find out if she is doing well at her job. She decides to ask the customers that seem really happy and enjoy her as a waitress to see how she is doing. What kind of sample is she obtaining?

A. Biased Sample

B. Unbiased Sample

C. Random Sample

D. Self-selected Sample

MM3D3

70. A new type of medicine is being studied. The scientists decide to test the new medicine on lab rats by giving a group of them the new medicine and giving the other group of lab rats a placebo pill. What kind of study is being observed?

A. Observational Study

B. Biased Study

C. Experimental Study

D. Marginal Study

MM3D3

71. What is the inverse of $\begin{bmatrix} 2 & 1 & -2 \\ 5 & 3 & 0 \\ 4 & 0 & 8 \end{bmatrix}$?

A. $\begin{bmatrix} 24 & -8 & 6 \\ -40 & 24 & -10 \\ -12 & 4 & 1 \end{bmatrix}$

B. $\dfrac{1}{8}\begin{bmatrix} 24 & -8 & 6 \\ -40 & 24 & -10 \\ -12 & 4 & 1 \end{bmatrix}$

C. $\dfrac{1}{32}\begin{bmatrix} 24 & -8 & 6 \\ -40 & 24 & -10 \\ -12 & 4 & 1 \end{bmatrix}$

D. $\dfrac{1}{16}\begin{bmatrix} 24 & -8 & 6 \\ -40 & 24 & -10 \\ -12 & 4 & 1 \end{bmatrix}$

MM3A4b

72. The matrix below represents the number of bags that three stores used during a particular day.

	Store A	Store B	Store C
paper bags	150	110	95
plastic bags	50	100	105

How many paper bags will Store A and Store B use in six months (180 days)?

A. 37,800

B. 46,800

C. 36,900

D. 73,800

MM3A5

Index